A Guide to Housing Rabbits

A Collection of Articles on the Construction and Layout of Rabbit Houses and Hutches

By

Various Authors

Contents

LOCATION OF THE RABBITRY

HE selection of a special section of the country for rabbit raising is not of paramount importance, for rabbits are successfully raised in practically all localities. If, however, the breeder is free to choose a location the following points may well be kept in mind:

It is advantageous to locate in a rabbit-raising community. There are many small problems in management that can not be discussed adequately in a general treatise on rabbit raising, but these may be solved through actual experience in one's own community. Then too, by raising a certain breed or grade of rabbit a whole community can build up a reputation for its product as a community that an individual could never hope for. When the buyers learn that the type or breed of rabbits that they desire is to be obtained in unlimited quantities in a certain section they will naturally turn their first. Furthermore such community

1

concentration tends to decrease the difficulties experienced by the small breeder in growing and marketing his stock.

The matter of transportation facilities must always be considered, especially ready access to the market and the type of rabbit it demands. Good roads are of inestimable importance for, among other things, they make it possible for the farmer to market his products at any and all times, thus enabling him to take advantage of favorable fluctuations in market prices.

In general it may be said that rabbits are grown successfully in practically every part of the country. Rabbitries are now established in small towns, in cities, in the vicinity of larger cities, in rural districts, and on poultry and general farms. (Figure 8).

PLAN OF THE RABBITRY

The land on which rabbit hutches and shelters are to be built should be well drained and reasonably level to facilitate construction. It may or may not be planted in fruit or shade trees, but some trees are desirable to provide shade in summer and to encourage a feeling of seclusion and security for the rabbits. The ideal location is a reasonably

FIG. 8. A BIRD'S-EYE VIEW OF A RABBITRY IN AN ORANGE GROVE IN CALIFORNIA

level, well drained piece of land, the slope being gentle to the south. Productivity of the soil is an important consideration if the rabbitry is to be large enough to warrant the home growing of all the feed required. Plans should be made at the outset for possible expansion of the rabbitry, the initial investment being in the nature of a first unit, so that time and money spent in the construction of additional units will not be wasted.

The plan of the rabbitry is of paramount importance to efficiency in operation. The cost of handling and feeding a hundred or more does can be reduced with sufficient foresight, and certain arrangements and types of construction are necessary to assist in making the business profitable. As much thought should be given to the structures and equipment as to feeding and breeding. (Figure 9).

Rabbits will stand both hot and cold weather provided they are given seasonal shelter from wind, rain, and direct sunlight. In the southern part of the United States, and in other places where the climate is mild, it is not necessary to erect a building in which to place the hutches. A roof constructed over them to give shade and keep out the rain will meet most of the requirements in this

FIG. 9. WELL CONSTRUCTED RABBITRIES. THE HUTCHES ARE
SO ARRANGED AS TO REDUCE TO A MINIMUM THE LABOR OF
FEEDING AND HANDLING; (*Above*) EXTERIOR VIEW; (*Below*)
INTERIOR VIEW

5

particular. In the North, however, a structure is required to protect the does and the young against winter weather.

EQUIPMENT

The chief structures in the rabbitry consist of pens, nest boxes, hutches, and shelters or houses in which the animals are protected from the elements. A cottage, feed house, and barn constitute desirable additional equipment for the larger rabbitries.

No definite type of house can be described to meet the conditions confronting producers in all parts of the country. The essentials in determining the kind to build are concerned with location, climate, extent of the business, and the amount of money to be invested. In establishing a rabbitry one must keep in mind construction and equipment that will facilitate handling a certain number of rabbits with a minimum of manual labor. Reducing the care required in feeding, handling, and breeding practices, as well as in cleaning the hutches and keeping the houses sanitary, is of the utmost importance, and is dependent upon the type of construction adopted.

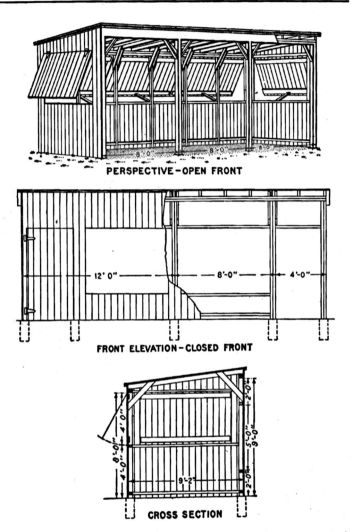

PERSPECTIVE-OPEN FRONT

FRONT ELEVATION-CLOSED FRONT

CROSS SECTION

FIG. 10. CONSTRUCTION DETAILS OF A RABBIT HOUSE

The foundation and floor should be of concrete, to expedite thorough cleaning. In some localities the ground may serve as the floor. There should be overhead ventilation, and windows should be so arranged that they may be opened to permit free circulation of air in warm weather. The windows should face the south or southeast to give the greatest benefit from its sunlight. An up-to-date poultry or hog house will furnish the correct idea for sunlight and ventilation. Water should be accessible to the rabbits, and arrangements should be made for washing feed dishes. Electric light facilitates the work of the producer in evenings and early winter mornings. Feed bins, scale, and a place to handle and examine the rabbits are also required. The kind and quantity of equipment necessary depends entirely upon the scope of the undertaking—the number of rabbits kept for breeding purposes.

MATERIAL REQUIRED FOR RABBIT HOUSE (FIG. 10).

1/3 cubic yard concrete for post footings.
8 2-inch by 4-inch by 9-foot posts.
9 2-inch by 4-inch by 8-foot posts.
18 2-inch by 4-inch by 8-foot plates and girts.
6 2-inch by 4-inch by 10-foot end girts.

13 2-inch by 4-inch by 10-foot rafters.
14 1-inch by 6-inch by 3-foot braces.
60 1-inch by 6-inch by 8-foot tongued and grooved sheathing (rear).
40 1-inch by 6-inch by 9-foot tongued and grooved sheathing (ends).
2 1-inch by 8-inch by 12-foot rear facia.
280 feet board measure roof sheathing.
70 linear feet 1-inch by 1-inch under edge of roof.
Prepared roofing for 260 square feet area.
6 pairs 6-inch T hinges.

For open-front shelter add the following:
6 2-inch by 4-inch by 3-foot braces.
2 1-inch by 8-inch by 12-foot facia.
14 1-inch by 4-inch by 9-foot post casings.

For closed shelter add the following (omit the three items under open front):
2 2-inch by 4-inch by 9-foot posts.
7 2-inch by 4-inch by 8-foot girts.
20 1-inch by 6-inch by 10-foot tongued and grooved sheathing.
32 1-inch by 6-inch by 5-foot tongued and grooved sheathing.
6 1-inch by 6-inch by 3-foot braces.
2 pairs 6-inch T hinges.

The feed house and barn when considered necessary may be a combination structure used to store grain, hay, and root crops and to mix rations. The sole purpose of this building is to provide a place for the storage and preparation of feed. It should, therefore, be used strictly as a feed house, and not as a hospital or quarantine pen for sick rabbits or a place to dress rabbits for the market. If there is no other convenient building for these purposes, one also should be provided, particularly on large rabbitries. (Figure 11).

The hutches designed to serve as isolation and hospital pens should be placed at least fifty feet from the breeding hutches. They should be so located that the men handling and feeding the regular stock need not come in contact with them.

Metal and crockery dishes are used for both feed and water. The crockery dishes that are to serve for a water supply should be of such shape that the rabbits can not tip them over. If the rabbit hutches are so constructed as to include a suitable manger for hay and a trough for grain feed it will not be necessary to place a dish for grain in the hutch.

MATERIALS REQUIRED FOR RABBIT-HUTCH
SHELTER

The following materials will be needed for the rabbit-hutch shelter shown in figures 12 and 13.

FIG. 11. ALL METAL HUTCHES FACILITATE FEEDING, WATER-
ING AND CLEANING AT THE UNITED STATES RABBIT EXPERI-
MENT STATION

5 sacks of cement, ⅜ cubic yard of sand, and ¾ cubic yard of gravel for 0.8 cubic yard of concrete for post footings. (Use 1 part portland cement to 2 parts of sand and 4 parts of gravel. In localities where frost action is a consideration the footings should be placed below the frost line and be 8 by 8 inches in cross section. If the footings are 8 by 8 inches by 2 feet 8 inches the following materials will be needed: 10 sacks of cement, ¾ cubic yard of sand, and 1½ cubic yards of gravel. Precast footings cost less than those cast in place.)

36 strap irons, ¼ by 1½ by 16 inches, one end drilled for two ⅜-inch lag screws.

72 lag screws, ⅜ by 4 inches.

18 posts, 4 by 4 inches by 10 feet.

18 posts, 4 by 4 inches by 8 feet.

3 cross ties, 2 by 4 inches by 14 feet.

12 cross ties, 2 by 4 inches by 12 feet.

7 longitudinal ties, 2 by 4 inches by 16 feet.

1 longitudinal tie, 2 by 4 inches by 8 feet.

90 knee braces, 2 by 4 inches by 2 feet.

18 purlins, 2 by 4 inches by 12 feet.

26 rafters, 2 by 4 inches by 16 feet.

13 rafters, 2 by 4 inches by 8 feet.

5 facias, 2 by 6 inches by 16 feet.

2 facias, 1 by 8 inches by 16 feet.

1 facia, 1 by 8 inches by 8 feet.

8 facias, 1 by 5 inches by 3 feet.

4 facias, 1 by 6 inches by 3 feet.

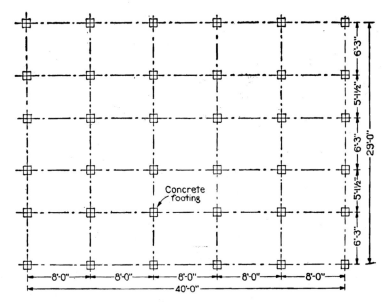

FOUNDATION PLAN

Concrete footing

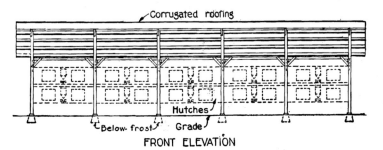

Corrugated roofing

Hutches

Below-frost   Grade

FRONT ELEVATION

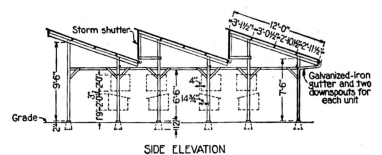

Storm shutter

Galvanized-iron gutter and two downspouts for each unit

Grade

SIDE ELEVATION

FIG. 12. CONSTRUCTION DETAILS FOR A RABBIT-HUTCH SHELTER

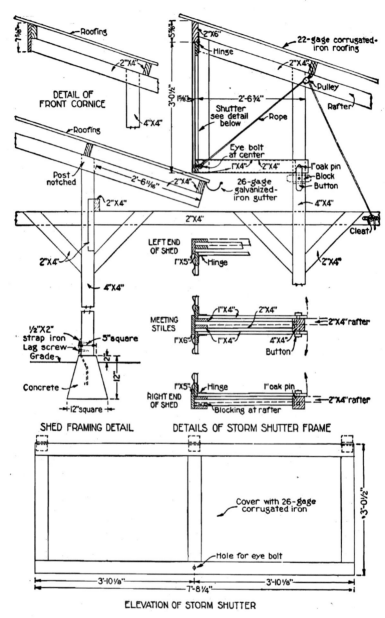

DETAIL OF FRONT CORNICE

Roofing

2"X4"

Roofing

Post notched

2'-6 11/16"

2"X4"

2"X4"

4"X4"

2"X4"

4"X4"

½"X2" strap iron
Lag screw
Grade

Concrete

5"square

12"square

SHED FRAMING DETAIL

2"X6"

Hinge

22-gage corrugated-iron roofing

2"X4"

Pulley

3'-0½"

2'-6¾"

Shutter
see detail
below

Rafter

Rope

Eye bolt
at center

1"X4"   2"X4"

26-gage
galvanized-
iron gutter

1"oak pin
Block
Button

4"X4"

2"X4"

Cleat

LEFT END OF SHED

1"X5"   Hinge

2"X4"

MEETING STILES

1"X4"   2"X4"

1"X6"   1"X4"   4"X4"

Button

2"X4"rafter

RIGHT END OF SHED

1"X5"   Hinge   1"oak pin

Blocking at rafter

2"X4"rafter

DETAILS OF STORM SHUTTER FRAME

Cover with 26-gage
corrugated iron

Hole for eye bolt

3'-10⅛"   3'-10⅛"

7'-8¼"

3'-0½"

ELEVATION OF STORM SHUTTER

FIG. 13. ADDITIONAL DETAILS FOR RABBIT-HUTCH SHELTER

14

12 struts, 2 by 4 inches by 2 feet 6 inches.
20 hangers, 1 by 4 inches by 2 feet 9 inches.
24 hangers, 1 by 4 inches by 3 feet 3 inches.
 7 blocks, 1 by 4 inches by 12 feet.
30 pieces, 1 by 5 inches by 8 feet—for shelter frame.
12 oak pins (1 inch), 7 inches long.
 1 piece, 1 by 2 inches by 10 feet—for buttons.
50 hinges, 3 by 3 inches, with screws.
20 round-head screws, 2½ inches long with washers—
    for holding buttons.
10 eyebolts, ⅜ by 1½ inches, 1-inch eye and nut.
10 screw eyes (1-inch eye) or heavy staples.
10 swivel pulleys for ¼-inch rope.
10 cleats (4 inches) for ¼-inch rope, with screws.
90 linear feet ¼-inch sash cord.
120 linear feet 4-inch gutter with hangers.
38 linear feet 3-inch downspout.
 6 turn-outs for 3-inch downspout.
12 bends (90°) for 3-inch downspout. (All gutters
    and spouts to be of 26-gage galvanized metal.)
40 pieces corrugated roofing 3 feet long, 26 inches
    wide—for shutters. (Plain galvanized metal,
    weather-resistant wall board, or ⅜-inch tongue-
    and-groove ceiling could be used in place of cor-
    rugated metal.)
66 pieces corrugated roofing, 7 feet long, 26 inches
    wide.
66 pieces corrugated roofing, 6 feet long, 26 inches
    wide. (Roofing to be 24-gage, 2½-inch cor-
    rugations, and heavily galvanized; nails for se-

curing roofing should be rust-resistant, providing
security against leaks.)

18 pounds sixteenpenny common nails.
2 pounds eightpenny common nails.
1 pound eightpenny finishing nails.
5 pounds tenpenny common nails.
10 gallons of paint for woodwork (three coats).

BREEDING HUTCHES

Numerous designs and styles of rabbit hutches
have been used but none of them as yet has proved
to be entirely satisfactory in every detail. Hutches
and nest boxes require frequent cleaning and spray-
ing to keep them free from vermin and to protect
the health of the animals. The construction, therefore, should be as simple as possible, and ample provision should be made for light and air. Both the
comfort of the animals and the convenience of the
caretaker should be kept in mind in planning the
construction.

Rabbit hutches should have about 10 or 12 square
feet of floor space. The standard hutch is 4 feet
long, 2½ feet deep, and 2 feet high. To economize
in space and to facilitate feeding and handling they
are frequently built two tiers high. If more tiers
are added, these are inconvenient to reach and hard

to clean, and it becomes difficult to observe the animals in them.

Rabbits are more easily cared for and are less likely to become diseased in well-built hutches of the size recommended than in poorly built temporary ones, for these latter become foul and unwholesome unless frequently cleaned and rebedded with straw, leaves, or other absorbents. Wire floored hutches require no bedding and are easily kept in good order.

A SATISFACTORY TYPE OF HUTCH

The many kinds of hutches now in use vary somewhat in general construction but involve the same fundamental ideas. All have either wire-mesh or slat-wood floors. The hutches shown in Plate 12, and in variations of this type, have given splendid service in facilitating feeding, watering, and handling. The floor is made of hardware cloth, to permit the droppings to fall to the ground. The hutch lends itself readily to tier construction, and may be built on legs or suspended from the rafters, or fastened to the uprights of the building. (Figures 14 and 15). This provides a clear floor space, and facilitates cleaning and flushing it.

FIG. 14. HUTCHES ARRANGED ON BOTH SIDES OF AN ALLEY
WAY IN CALIFORNIA ARE PROTECTED FROM THE DIRECT
RAYS OF THE SUN BY A LATTICE SUPERSTRUCTURE

18

The construction of a U-shaped hay rack and a drawer grain trough cuts down the labor of feeding and lessens waste of feed. Young rabbits will crawl into the trough, but this can be prevented by attaching cross-strips of galvanized iron 2-inches wide to a strip of wood extending just above the trough. The iron strips should be placed at right angles to the wood and reach to the end of the trough. To feed the rabbits the hay is placed in the hay rack, and the drawer is pulled out and the grain placed in it.

An especially desirable feature of this type of hutch (Figure 15) is the galvanized iron covering over the lower tier. The droppings and urine that pass through the wire mesh floor of the top hutch fall to the galvanized iron roof of the lower and then to the floor or to the ground. The galvanized wire mesh is so arranged in constructing the hutch that it completely covers all the wood work, leaving none exposed for the rabbits to gnaw. The doors are equipped with hinges made of nails and staples and an easily closing catch to keep them tightly closed. Water is best supplied to the rabbits in the hutch in earthern crocks or galvanized iron pans. Automatic water fonts are satisfactory if properly installed. The system should not per-

FIG. 15. A TYPE OF HUTCH THAT FACILITATES FEEDING, WATERING AND CLEANING. IT IS ALSO WELL ADAPTED TO TIER CONSTRUCTION; THE ARRANGEMENT OF ONE HUTCH OVER THE OTHER

mit water to run from one hutch to another, as this may spread disease. Each watering device in a hutch should be a separate unit.

Details of construction of the hutch, and of units composed of two and four hutches, are shown in Figs. 16 and 17.

MATERIAL FOR 2-HUTCH UNIT (FIG. 16).

4 2-inch by 2-inch by 5-foot posts.
2 1-inch by 2-inch by 6-foot knee braces.
2 1-inch by 3-inch by 8-foot bottom frames.
2 1-inch by 3-inch by 3-foot bottom frames.
1 1-inch by 4-inch by 8-foot bottom frames.
2 1-inch by 4-inch by 8-foot top frames.
1 1-inch by 3-inch by 8-foot top frame.
2 1-inch by 4-inch by 3-foot top frames.
8 $7/8$-inch by $5\frac{1}{2}$-inch by 8-foot tongued and grooved sheathing.
2 1-inch by 2-inch by 8-foot doors.
1 1-inch by 3-inch by 5-foot jambs.
1 1-inch by 2-inch by 12-foot manger.
1 1-inch by 3-inch by $1\frac{1}{2}$-foot manger.
1 1-inch by 4-inch by 2-foot manger.
1 1-inch by 6-inch by 3-foot manger.
1 1-inch by 12-inch by 6-foot manger.
27 linear feet, 1-inch mesh poultry wire, 2 feet wide.
1 piece $5/8$-inch mesh hardware cloth, 2 feet 6 inches by 7 feet 6 inches.
Roll roofing, 3 feet by 8 feet.

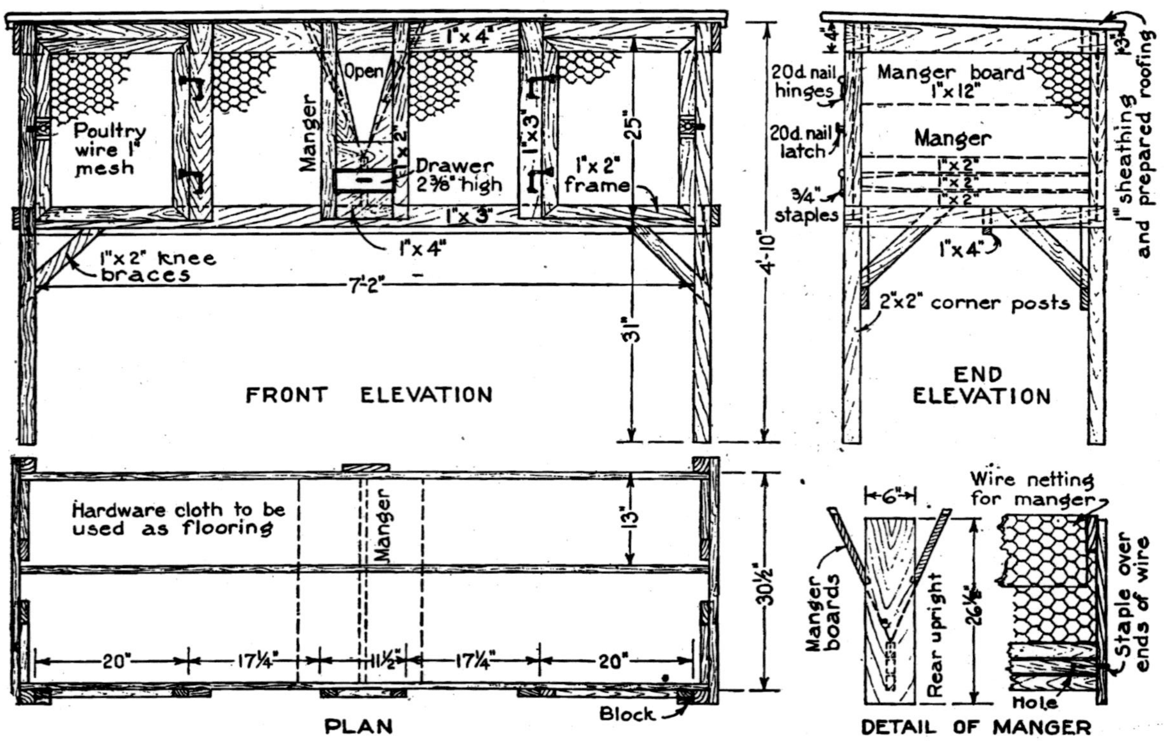

END ELEVATION

1" sheathing and prepared roofing

Manger board 1"x 12"

Manger

1"x 4"

2"x 2" corner posts

20d nail hinges

20d nail latch

3/4" staples

Staple over ends of wire

Wire netting for manger

Hole

Rear upright

26 1/4"

6"

Manger boards

DETAIL OF MANGER

FRONT ELEVATION

4'-10"

25"

31"

7'-2"

1"x 2" frame

1"x 3"

1"x 4"

Drawer 2 5/8" high

1"x 3"

1"x 4"

Open

Manger

Poultry wire 1" mesh

1"x 2" knee braces

30 1/2"

13"

20"

17 1/4"

11 1/2"

17 1/4"

20"

Block

Manger

Hardware cloth to be used as flooring

PLAN

FIG. 16. DETAILS OF CONSTRUCTION OF A TWO-HUTCH UNIT

22

MATERIAL FOR 4-UNIT HUTCH (FIG. 17).

4 2-inch by 2-inch by 6-foot posts.
4 1-inch by 3-inch by 8-foot lower and upper frames.
4 1-inch by 3-inch by 3-foot lower and upper frames.
2 1-inch by 4-inch by 8-foot lower and upper frames.
1 1-inch by 6-inch by 8-foot dropping pan frame.
1 1-inch by 4-inch by 8-foot dropping pan frame.
2 1-inch by 4-inch by 3-foot dropping pan frames.
1 1-inch by 4-inch by 3-foot (makes 2 1-inch by 16-inch by 3½-inch by 2-inch).
1 1-inch by 4-inch by 8-foot roof frame.
1 1-inch by 3-inch by 8-foot roof frame.
2 1-inch by 4-inch by 3-foot roof frames.
3 1-inch by 2-inch by 12-foot doors.
1 1-inch by 3-inch by 10-foot jambs.
1 1-inch by 2-inch by 10-foot manger.
1 1-inch by 6-inch by 4-foot manger.
4 1-inch by 4-inch by 3-foot rails.
1 1-inch by 3-inch by 3-foot rail.
1 1-inch by 12-inch by 3-foot rail.
6 linear feet, 1-inch mesh poultry wire, 4 feet wide.
20 linear feet, 1-inch mesh poultry wire, 2 feet wide.
2 pieces ⅝-inch mesh hardware cloth, 7 feet 6 inches by 2 feet 6 inches.
1 piece No. 20-gauge galvanized iron, 3 feet by 9 feet.
1 piece No. 16-gauge galvanized iron, 3 feet 4 inches by 8 feet.

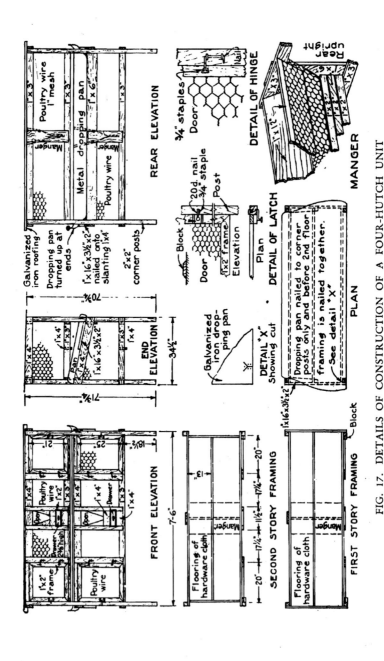

FIG. 17. DETAILS OF CONSTRUCTION OF A FOUR-HUTCH UNIT

24

All-Metal Sanitary Hutches

Rabbits sometimes do considerable damage to hutches made of wood by gnawing. The fact that these hutches are also difficult to keep clean and sanitary has led to the development of the all metal hutch. The material and the method of construction prevent the rabbits from damaging these hutches. They can be thoroughly cleaned and disinfected, assuring good sanitation at all times. The hay racks and the feed troughs can be filled without opening the hutch door or disturbing the rabbits. It costs practically no more to install such hutches ready-built than to buy materials and have them constructed. This type of hutch is comparatively new equipment for the rabbit industry, and future developments alone will determine the relative degree of its usefulness. (Figure 18).

Convenient, two-compartment hutches, as illustrated in figures 16 and 17, can be made for the most part of electro spot-welded wire fabric, 1- by 2-inch mesh. In 24-inch widths this material can be used for the sides and ends. Labor can be saved by using one length of the wire fabric for all sides, bending it at the corners; but if separate pieces are cut for front, back, and sides, these can be fastened

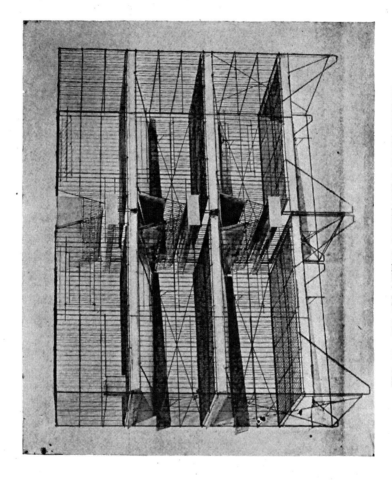

FIG. 18. AN ALL-METAL SANITARY RABBIT HUTCH

together at the corners with spiral wire or a soft galvanized stovepipe wire of about 12 gage. In the same way the sides and ends after assembling can be attached to a bottom of either galvanized ⅝-inch-mesh hardware cloth or perforated metal sheets (Figure 16). If a sloping roof is used on the hutch, as is required for lower sections in tier installation, the end sections are bent to the desired pitch, and this also provides additional rigidity. The top can be made of either electro spot-welded wire fabric or galvanized sheet iron. The hay manger between the compartments can be constructed of 1-inch-mesh, 16-gage poultry netting, as shown in figure 16. The electro spot-welded wire fabric can be cut as desired to provide openings for doors and feed trough and permit access to the hay manger. Methods of constructing a wire lock or fastener, hinges, and partitions beneath the hay manger. (Figure 17).

MATERIAL REQUIRED FOR ALL-METAL HUTCH

As shown in figures 19 and 20, two all-metal hutches (one upper and one lower) can be constructed of the following materials:

Wire fabric.—All wire fabric should be electro spot-welded, 12½-gage, 2-inch vertical mesh, and 1-inch

horizontal mesh. The following quantities are required:
    48 linear feet, 24 inches wide—for walls and four
        doors.
    8 linear feet, 30 inches wide—for top of upper
        hutch.

Galvanized sheet steel.—Wherever practicable the edges of the galvanized sheet steel used should be turned under ¼ inch for stiffness. The following quantities are required:

    1 sheet, 26 gage, 4 by 8 feet, cut into the follow-
        ing pieces:
        4 pieces, 15 inches by 2 feet 8 inches—for
            sides of hay manger.
        2 pieces, 16 inches by 2 feet 8 inches—for
            center of hay manger.
        4 pieces, 1½ by 18 inches—for crock straps.
        4 pieces, 2 inches by 2 feet 6 inches—for
            manger mesh guard.
        4 pieces, 7 by 5 inches—for breeding-record
            cards.
        4 pieces, 3 by 5 inches—for feed-record
            cards.
        2 pieces, 6 inches by 2 feet 6 inches—for
            feed-trough division.
        2 pieces, 12 inches by 2 feet 6 inches—for
            trough guides.
    1 sheet, 24 gage, 3 by 3 feet, cut into the follow-
        ing pieces:
        2 pieces, 4 by 9 inches—for front of feed
            trough.

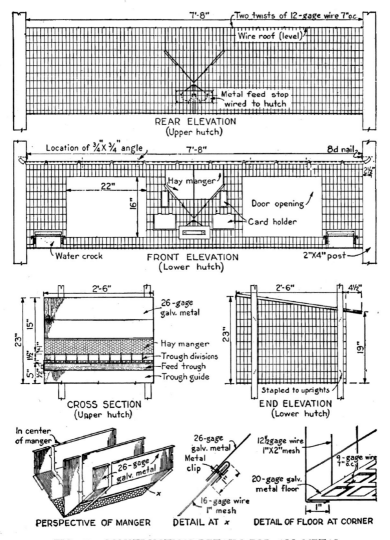

7'-8" — Two twists of 12-gage wire 7"o.c.

Wire roof (level)

Metal feed stop
wired to hutch

REAR ELEVATION
(Upper hutch)

Location of ¾ X ¾" angle — **7'-8"** — 8d nail

Hay manger

22"

16"

Door opening

Card holder

Water crock

FRONT ELEVATION
(Lower hutch)

2"X4" post

2'-6"

26-gage
galv. metal

23" 15"

1½" ¾"

5" ½"

Hay manger
Trough divisions
Feed trough
Trough guide

CROSS SECTION
(Upper hutch)

2'-6" 4½"

23" 19"

Stapled to uprights

END ELEVATION
(Lower hutch)

In center
of manger

26-gage
galv. metal

x

PERSPECTIVE OF MANGER

26-gage
galv. metal
Metal
clip

16-gage wire
1" mesh

DETAIL AT x

12½gage wire
1"X2"mesh

9-gage wire
7" o.c.

20-gage galv.
metal floor

1"

DETAIL OF FLOOR AT CORNER

FIG. 19. CONSTRUCTION DETAILS FOR ALL-METAL
RABBIT HUTCH

29

2 pieces, 1½ by 10½ inches—for feed-
handles.

2 pieces, 1 foot 3 inches by 2 feet 7 inches—
for feed trough.

1 sheet, 26 gage, 3 by 8 feet—for roof. (Not
cut into pieces.)

4 angles, 20-gage galvanized iron, ¾-inch sides,
31 inches long.

Double each leg—for roof support of lower
hutch at spaced intervals.

Cut away enough of one side at both ends
for the other side of angle to rest on sides
of hutch. Riveting angle in place to
under side of roof gives added rigidity.

2 sheets, 20 gage, 2 feet 6 inches wide by 7 feet
8 inches long—for floors.

Perforate with ⅝-inch square holes ⅞ inch
on centers both ways.

Wire:

5 linear feet, 12-gage galvanized soft wire—for
lacing.

7 linear feet, 9-gage copper wire—for lacing.

2 linear feet, 8-gage copper wire—for lacing.

4 pieces, 13 inches long, No. 9 soft galvanized
wire—for crock hooks.

Miscellaneous:

1 piece, 1-inch-mesh, 16-gage poultry netting 30
by 12 inches—for manger.

2 pieces strap iron, ¼ by ¾ inch by 30 inches—
for feed-trough division.

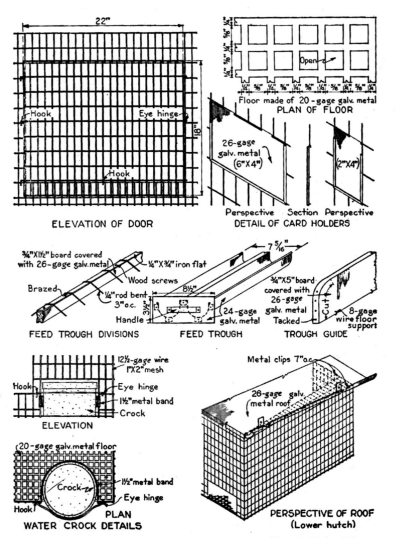

ELEVATION OF DOOR

PLAN OF FLOOR

DETAIL OF CARD HOLDERS

FEED TROUGH DIVISIONS

FEED TROUGH

TROUGH GUIDE

ELEVATION

WATER CROCK DETAILS

PERSPECTIVE OF ROOF
(Lower hutch)

FIG. 20. ADDITIONAL CONSTRUCTION DETAILS FOR ALL-METAL RABBIT HUTCH

18 pieces, ¼-inch-diameter, iron rods, 10 inches long—for feed-trough division.

6 round-head wood screws, ⁵⁄₁₆ by 1 inch—for feed-trough division.

16 large carpet tacks—for trough guide.

53 brass clips or cotter pins, ⅛ by ⁹⁄₁₆ inch, with flat heads (see detail x fig. 19).

4 common eightpenny wire nails.

16 staples, ¾ inch--9 gage—for fastening to support.

2 pieces, wood boards (yellow pine or other kind), ¾ by 1½ inches by 2 feet 6 inches—for feed-trough division.

2 pieces, wood boards (yellow pine or other kind), ¾ by 5 inches by 2 feet 6 inches—for trough guide.

4 earthenware crocks, 7 inches in diameter, 4 inches deep.

The Nest Box

Formerly it was a common practice to improvise an inexpensive nest box from a packing box. This type of nest box, however, has proved to be more expensive when in actual use than one made especially for the purpose. No top was provided and, consequently, the doe in hopping into the nest would frequently tramp on and injure her young. Cracks developed in the bottom of some of these

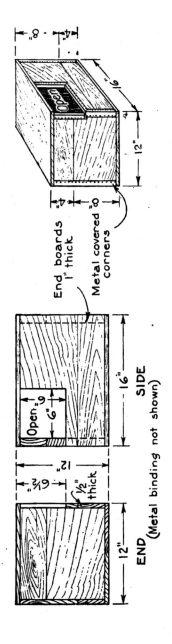

FIG. 21. DETAILS OF CONSTRUCTION OF A WOODEN NEST BOX

33

boxes of sufficient size to permit the foot of a young rabbit to slip through, resulting in a broken leg. Not only is a substantial nest box designed especially to accommodate the doe and her litter desirable, but it is necessary, as it will be the home of each family of young rabbits and the doe for some time. Nest boxes should be large enough to prevent crowding but small enough to keep doe and litter warm. Two kinds of boxes are used extensively—the box type (fig. 21) and the nail keg nest box (fig. 22). The box type is constructed so the top and bottom can be removed to facilitate cleaning.

The nail keg-nest is inexpensive and easy to construct. A nail keg with metal end and hoops is best for the purpose. One with a head diameter of 13 inches is preferable for does weighing more than 12 pounds; a diameter of 11½ inches for those weighing 8 to 12 pounds; and a diameter of 10 inches for those weighing less than 8 pounds.

MATERIAL FOR NEST BOX (FIG. 22)

2 ½-inch by 12-inch by 16-inch.
2 ½-inch by 7½-inch by 16-inch.
2 ½-inch by 3½-inch by 16-inch.
2 1-inch by 7½-inch by 11-inch.

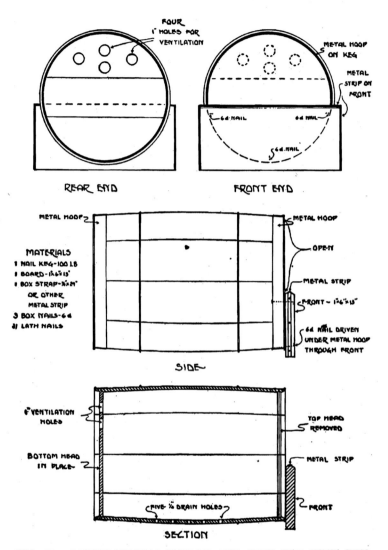

FOUR
1" HOLES FOR
VENTILATION

METAL HOOP
ON KEG

METAL
STRIP ON
FRONT

6d. NAIL   6d NAIL

6d. NAIL

REAR END

FRONT END

METAL HOOP

METAL HOOP

OPEN

METAL STRIP

FRONT - 1"x6"x13"

6d NAIL DRIVEN
UNDER METAL HOOP
THROUGH FRONT

MATERIALS
1 NAIL KEG-100 LB
1 BOARD-1"x6"x13"
1 BOX STRAP-¾"x24"
OR OTHER
METAL STRIP
3 BOX NAILS-6d
8 LATH NAILS

SIDE

1" VENTILATION
HOLES

TOP HEAD
REMOVED

BOTTOM HEAD
IN PLACE

METAL STRIP

FIVE ¼" DRAIN HOLES

FRONT

SECTION

FIG. 22. CONSTRUCTION DETAILS FOR NAIL-KEG NEST BOX

2 1-inch by 3½-inch by 11-inch.
5 linear feet No. 20-gage metal
ribbon, 1½ inches wide.

SELF-FEEDERS

The self-feeding system is primarily adapted to raising rabbits for market. It is not recommended for feeding dry does or herd bucks, or for developing breeding stock. The full feeding of these classes of rabbits would result in their attaining a higher degree of condition than is desirable for breeding.

The does should be carefully watched the first few days after kindling. If the size of the litter is materially reduced for any reason, or if the doe produces more milk than the young will consume, it may be necessary to adjust the size of the litter by transferring young from another litter or to restrict the doe's ration for a few days to check the heavy milk secretion and avoid udder complications that might follow.

The self-feeder is not adapted for use with a mixed ration that can be separated by the rabbits, as the animals in their search for the more palatable kinds of feed will scratch out and waste considerable of the ration. On the other hand, when the

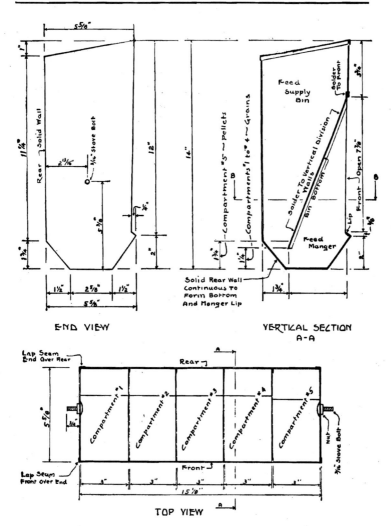

FIG. 23. CONSTRUCTION DETAILS FOR AN ALL-METAL RABBIT SELF-FEEDER. END, VERTICAL AND TOP VIEWS

different grains and protein supplements are placed in separate compartments, so that the rabbits have free access to any kind of feed at all times, they will consume the particular feed they desire and the quantity wasted will be negligible.

ADVANTAGES

The advantages of the self-feeding system are:

1. It has proved satisfactory in developing fryers and roasters for market.
2. It prevents waste and contamination and requires less feed than the hand-feeding system to produce a unit of gain in live weight.
3. It saves much labor and makes possible a consistently high-quality market product.
4. It prevents inefficiency and carelessness in feeding during the finishing period.
5. Self-fed rabbits gain more rapidly in weight than hand-fed rabbits.
6. Self-fed rabbits attain a high degree of finish—fryers give a carcass yield of 50 to 57 percent of their live weight and roasters 60 to 65 percent.

KINDS OF SELF-FEEDERS

Two general kinds of self-feeders are illustrated —the all-metal (figs. 23, 24, and 25) and the 5-

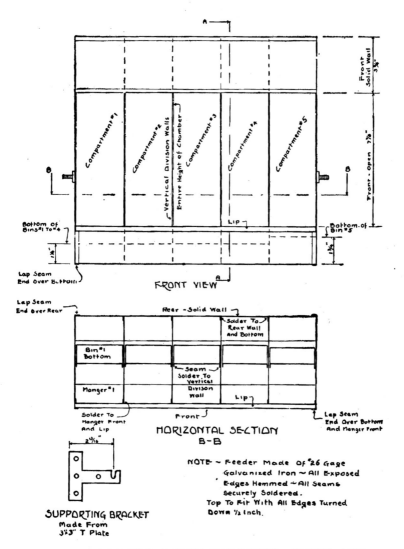

FIG. 24. ALL-METAL RABBIT SELF-FEEDER. FRONT AND
HORIZONTAL VIEWS

gallon, oil-can feeders (figs. 26 and 27) of such capacity that when filled about twice a week they will furnish a doe and her litter with all the concentrates required. The angles of the sides and the size of the throat should be as indicated in the drawings; otherwise, the feed may move too freely and be wasted, or it may choke the hoppers and become unavailable.

The all-metal self-feeder is designed especially for experimental work where it is necessary to keep accurate records. It has the added advantage of not requiring floor space and can be constructed by a tinsmith. Figures 23, 24, and 25 show the materials required and the method of construction and attachment.

The oil-can feeder is less expensive, can be made by anyone with mechanical ability, and is satisfactory for use in commercial rabbitries. Two types are illustrated in figures 26 and 27. One (fig. 26) is constructed from a 5-gallon oil can; the side of an apple box, ⅜-inch plywood, or other wood material of similar thickness; two strips of 26-gage galvanized iron, 30 inches long; and one piece of 26-gage galvanized iron 11½ inches square. The other type (fig. 27) is similar but lacks the galvanized iron parts used in type 1. The first is more

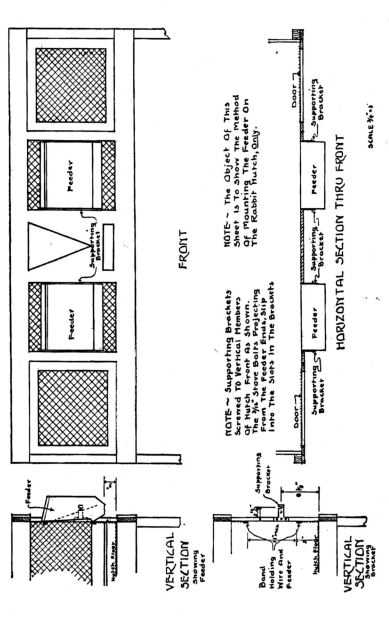

FRONT

NOTE:~ The Object Of This Sheet Is To Show The Method Of Mounting The Feeder On The Rabbit Hutch, Only.

NOTE:~ Supporting Brackets Screwed To Vertical Members Of Hutch Front As Shown. The 3/16" Stove Bolts Projecting From The Feeder Ends, Slip Into The Slots In The Brackets

VERTICAL SECTION
Showing Feeder

Feeder

VERTICAL SECTION
Showing Bracket

Band Holding Wire And Feeder

Supporting Bracket

HORIZONTAL SECTION THRU FRONT

SCALE 1/2"=1'

FIG. 25. ALL-METAL RABBIT SELF-FEEDER. DETAILS OF INSTALLATION

41

easily constructed but the second is less expensive. In making either, it will be desirable to check the outside measurements of the oil can used; the detailed suggestions made are for a can 9⅜ inches wide by 13¾ inches high.

Directions for Making Oil-Can Self-Feeder

Type 1.—Cut the top out of a 5-gallon oil can and clean the can thoroughly. Cut two openings in both front and rear (see "front-rear," fig. 26). These should each be 4 inches from the bottom and, when finished, should be 4 inches high by 3 inches wide. The openings, as cut, should be smaller than when finished to allow for a ¼-inch hem to be turned on the inside of the can to make a smooth edge. (See "front-rear" and "cut openings in can," fig. 26).

Use a side of an apple box or ⅜-inch plywood, or other wood material of the same thickness, for the partitions (S and T). Make one main partition (S) and two side partitions (T).

Cut two bin bottoms (U) out of 26-gage galvanized iron. The strips should each be 30 inches long and 4⁷⁄₁₆ inches wide. An opening 7⅛ inches

42

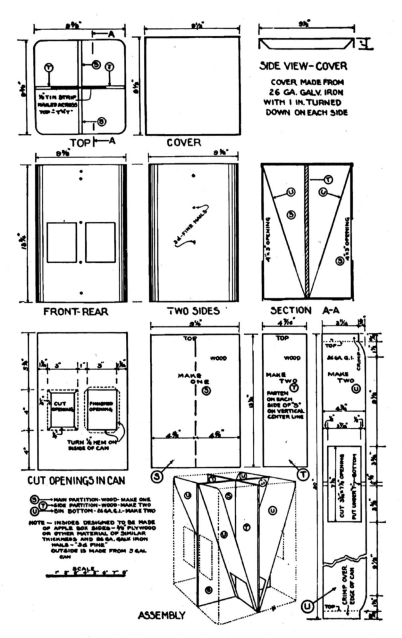

FIG. 26. SELF-FEEDER MADE FROM 5-GALLON OIL CAN, ⅜ INCH WOOD MATERIAL, AND GALVANIZED IRON

43

long and $3\frac{7}{16}$ inches wide should be centered 15 inches from one end of the strip and $\frac{1}{2}$ inch from each side.

The two top outside corners of each strip (U) should be cut out $\frac{3}{4}$ inch by $1\frac{3}{16}$ inch and the projecting shoulder rounded off $1\frac{1}{2}$ inches to permit the strips (U) to fit snugly into the rounded corner of the can.

In assembling the self-feeder three threepenny fine nails should be equally spaced on the center line of the main partition (S) and driven through it, after which it is turned over and placed on one of the side partitions (T) that is standing on edge. The partition (T) is then securely fastened to the median line of S by three more nails. The other side partition (T) is then fastened to S by driving it down on the three protruding nails.

Fasten a $\frac{1}{8}$-inch tin strip across the top edge of the two partitions (T) (see "top," fig. 26) to prevent spreading. The assembled pieces S and T should be placed on a flat surface, top down, and the bin bottoms (U) centered on the bottom of side partition (T), being sure that the rounded corners are opposite the main partition (S). Secure one bin bottom (U) to one partition (T) with two threepenny fine nails; then center and fasten the

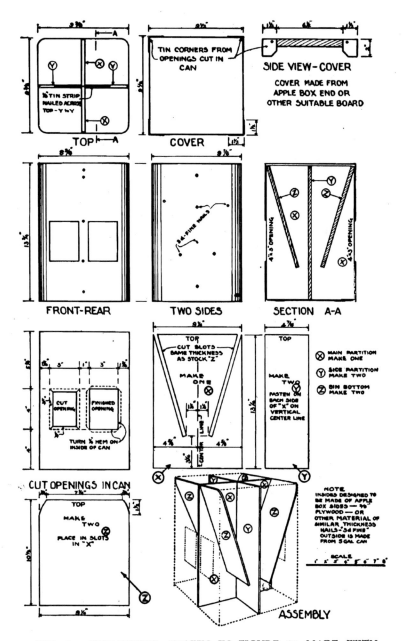

FIG. 27. SELF-FEEDER SHOWN IN FIGURE 25 MADE WITH
WOOD INSTEAD OF GALVANIZED IRON

45

other bin bottom (U) to the opposite side partition (T) in like manner.

The assembled parts, (S, T, and U) (see "assembly," fig. 26) are now ready to be placed in the 5-gallon can and are fastened to the can with three-penny fine nails, as shown on the "front-rear" and "two sides" drawings. Bend and crimp top of U over edge of can to secure bin bottoms. Tack the can to the bottom of S and T to prevent sagging.

A piece of 26-gage galvanized iron 11½ inches square should have the corners cut and 1 inch turned down on each side to form a cover. (See "side view-cover," fig. 26.)

Type 2.—Cut the top out of a 5-gallon oil can and clean the can thoroughly. Cut two openings in both front and rear of the can. (See "front-rear" in drawing, fig. 27.) These should each be 4 inches from the bottom and, when finished, should be 4 inches high and 3 inches wide. The openings, as cut, should be smaller than the finished openings to allow for a ¼-inch hem to be turned on the inside of the can to make a smooth edge. (See drawing of cut and finished openings, fig. 27.)

Use a side of an apple box or ⅜-inch plywood, or other wood material of the same thickness, for parts X, Y, and Z. Make one main partition (X),

two side partition (Y), and two bin bottoms (Z).
Cut two slots in main partition (X) to receive the
two bin bottoms (Z), and fasten the two side partitions (Y) one on each side of X on the vertical
center to form four compartments.

In assembling X, Y, and Z three threepenny fine
nails should be equally spaced on the center line of
the main partition (X) and driven through it, after
which it is turned over and placed on one of the
side partitions (Y) that is standing on edge. The
partition (Y) is then securely fastened to the median line of X by three more nails. The other side
partition (Y) is then fastened to X by driving it
down on the three protruding nails.

Fasten a $\frac{1}{8}$-inch tin strip across the top edge of
the two partitions (Y) (see "top," fig. 27) to prevent spreading.

Center the two bin bottoms (Z) in the slots in X
and secure with threepenny fine nails.

The assembled parts (X, Y, and Z) (see "assembly" fig. 27) are now ready to be placed in the can
and are secured to it with threepenny fine nails, as
shown on the "front-rear" and "two sides" drawings (fig. 27). Tack the can to the bottom of X
and Y to prevent sagging.

The cover, 9½ inches square, is made from an end board of an apple box. Four tins, each 2 by 3 inches, for securing the corners are made from openings cut in the can. (See "side view-cover," fig. 27.)

HUTCHES

It is most difficult to advise on definite methods of housing rabbits, as so much depends on what materials are available.

For small domestic breeders, unless money is no object, make your own. It is not necessary to have any previous experience of carpentering.

The following hints and accompanying diagrams should enable you to produce a very satisfactory result. For many generations, cottage people have kept rabbits with every success in what would now be termed most unsuitable conditions, and frequently a doe and 8 youngsters have thriven in a hutch 18 in. by 12 in., literally sitting on top of each other, with cleaning out operations unknown. To add to the discomfort this equipment was often placed in a damp cellar without any ventilation.

This is mentioned to show that rabbits are hardy and can be made to thrive in most difficult circumstances.

Before deciding on the type of hutch to make, arrange where to keep the rabbits.

For the small householder with a small space, the choice is limited, and usually the breeding does are hutched in a small outhouse, with additional running-on hutches beside the garden wall or fence.

When planning this make sure that the shed is well ventilated and the outside site, if walled in, is not too hot.

It is a mistake to place hutches facing the sun in the summer months; the heat inside would be terrific. In the winter and early spring sun is beneficial to all ages.

If a brick wall or other similar type of fence divides the garden, a good plan is to place the hutches back to this, with a roof, covering the hutches, similar to the photograph reproduced in this book. (*See Plate* 46.)

It is always advisable to make hutches in separate units. Hutches built into the side of a shed, using the shed as part of the hutch, are unsuitable.

Hutches made in this way are difficult to clean, and it is difficult to keep rats out.

Also rabbits will gnaw any protruding edges and may seriously damage the building.

Separate units if stood a few inches away from the side of a building should be safe from rats.

For that reason keep the top of the hutch and floor clear of rubbish. Rats usually work from concealed places.

There are various ways and methods of keeping rabbits, depending on the ultimate aim, and facilities at hand.

For the ordinary small breeder whose interest is mainly

to grow a few rabbits for home consumption, only a small equipment and a moderate outlay is necessary.

For the beginner starting with 3 does and 1 buck, the first outlay should be for 1 stack of 3 compartments breeding hutches, and 1 single buck hutch. Plates 42 and 43 show suitable hutches for the purpose and sketches show how to make them.

The next step is to make provision for the increase in stock.

Note that each mother should be having a litter which should average 6 youngsters for weaning every 10 to 12 weeks. See diagram of annual output for a doe on page 109.

Assuming that the progeny will be killed for table at 4 to 5 months old, giving the producer 1 to 2 rabbits per week for his table, accommodation will be required for 4 litters of different ages at the same time (assuming does are mated at different stages).

It is not advisable to run rabbits of vastly different ages together.

Consequently for this purpose, it is advisable to use either the following: Morant type, see Fig. 32, Weaning type, see growing-on type, Plate 43, also Plate 46, and for the larger breeder the colony system, see Figs. 33 and 33a, pages 131–132. Further details of these types are given later.

Many breeders retard the progress of their stock by not making provision in advance and leaving the litters over-crowded with the mother.

This may cause permanent damage to the health of the youngsters. Any check with young growing stock is difficult to put right, and deaths at a later stage may be attributed to other faults. Also, running with the mother too long lessens the annual output from the breeding doe.

Remember when constructing a hutch that it is a lasting job that is needed, and every attention must be given to make the hutch water-tight. Paint or solignum is the best preservative for the outside and tar for the roof. Avoid tarring any part which is handled. In hot weather tar melts and becomes a nuisance.

Ruberoid roofing felt or galvanised corrugated iron are both suitable for roofing. If neither are procurable, a sound roofing can be made by first tarring the surface which it is required to cover, then placing ordinary sacking on, and tar and sand. Half inch mesh wire netting should be used for all doors, in as heavy gauge as possible. Always put the netting inside the door to prevent gnawing. Tools necessary for making hutches: saw, hammer and screwdriver and wirecutters.

When constructing a hutch every attention should be given to labour saving, hygiene and lasting wear. The sketch given overleaf is the result of many years' experiments, and has proved to be the most suitable in every respect for breeding does.

There is a lot of liquid coming from rabbits, and unless this can be cleaned out at frequent intervals the hutches will smell and also be injurious to the stock. The best means to avoid this is to adopt the drainage system as shown on sketch.

This has the advantage of allowing all liquid to drain from the hutch immediately, leaving the floor dry and free from smell.

If the hutch is outside, the drainage can be allowed to drop directly on the ground, if in an inside shed a tin guttering should be used to divert the drainage outside.

It is a saving to make the hutches in stacks of 3, which means that 3 separate hutches are made on the same framing, but having individual doors.

When possible construct hutches with $\frac{1}{2}$ or $\frac{5}{8}$ inch boards, matching for preference, on 2-inch by 1-inch quartering.

The wire netting door should be made of $\frac{1}{2}$-inch or $\frac{3}{4}$-inch

FIG. 27. DRAINAGE SYSTEM FOR HUTCHES
Side elevation of stack hutch 3 tier high

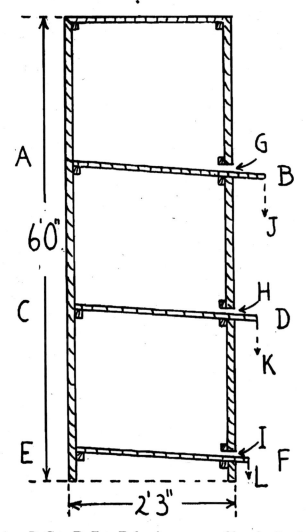

Floors. A to B, C to D, E to F slanting to rear of hutch allowing a fall of $1\frac{1}{2}$ inches.
At G, H and I leave an opening 1 inch at rear of hutch to keep space open for drainage.

Overlap floors at rear, allowing drainage to drop clear of hutch. Arrange protrusions as follows:

Distance from back of hutch at B 3 inches, D 2 inches, F 1 inch. Note hutch must be level, to allow dripping at J, K, L to fall clear of hutch.

wire netting of a heavy gauge, as rabbits soon bite through a thin gauge.

Always nail the wire netting inside the door. This prevents the rabbits gnawing the wood on the inside.

Remember that moisture from rain and inside drainage' will swell the wood. Consequently allowance must be made for this when constructing the doors by allowing a good space between the joints, ⅛ to ¼ in.

Hutch doors are opened several times each day and if difficult to manage the stock will get neglected.

BREEDING HUTCH

A good size for a breeding hutch is 3 ft. 6 in. long, 2 ft. 3 in. wide by 1 ft. 10 in. high.

By having one door to open for the whole front and a small wire netting door hinged on the outside door opening up half the front, it gives the valuable advantage of being able to open the whole front for cleaning out. (Fig. 27a, also Plate 42.)

Actually it is one door made on the other. The door going the whole length of the hutch, makes the cleaning-out operation easy. The wire door is used for feeding, etc. Rabbits generally have their litters in the darkest corner of the hutch, which in this hutch is behind the part not opened for feeding.

Nest boxes are sometimes used in this type of hutch, particulars and advantages of nest boxes will be found in this book under NEST BOXES (Fig. 34). These kind of hutches are used for inside use; if used for outside they should have a special roofing protection as shown in the photograph of running-on hutches, with roofing. Plate 46.

CONVERTING PACKING CASES

Packing cases are easily converted into suitable hutches.

The sketch on page 128 shows a simple and satisfactory way. Note the instructions and system adopted for fixing the door, erecting roof and legs.

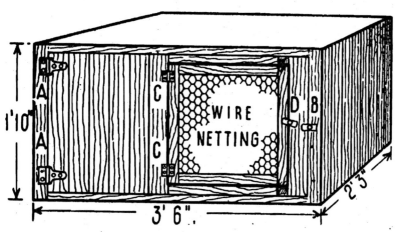

FIG. 27A. BREEDING HUTCH

WIDTH, 2 feet 3 inches.
HEIGHT, 1 foot 10 inches.
LENGTH, 3 feet 6 inches.
One door for total length of hutch, wire netting door hung on the main door.
Other details of construction in chapter on hutches. See special sketch for drainage.
Full length door. Hinged at A, buttons at B.
Small door hinge at C and button at D.
Small wire netting door used for feeding. Full length door for cleaning out.

(See plates of breeding hutches.)

Invariably the inexperienced fix doors with leather hinges, and fastening with bent nails, and doors which are difficult to open or shut. It takes little longer to make a good lasting job, so take the extra trouble.

Badly constructed hutches are not necessary and are time wasted, and prove a constant nuisance and impede proper supervision of stock. Many types of packing cases may be converted into hutches as shown, i.e. tea chests, sugar boxes,

BUCK HUTCH

Stud bucks are matured and need less exercise. Only a small hutch is required; 18 in. square is large enough. It is best to have this hutch away from the breeding does.

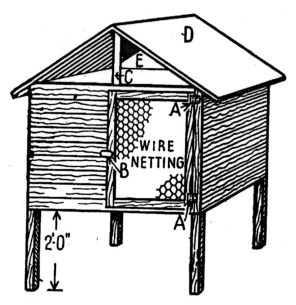

FIG. 28. PACKING CASE CONVERTED INTO HUTCH

Nail 4 uprights at the 4 corners of packing case, 2 inches by 1 inch material. Erect 2 feet from the ground.
A Hinges flat on door and *front* of hutch.
B Button.
C Upright centre front and back, supporting batten for nailing boards for roof D.
D Roof. Plain boards or matching $\frac{1}{2}$ inch by $\frac{3}{8}$ inch. Cover with roofing felt.
E Open space on top of hutch, suitable for storing food in dry.
DOOR, 2 inches by $\frac{3}{4}$ inch framing. $\frac{3}{4}$ inch mesh wire netting nailed on inside of door.
Size according to packing case. Suitable for breeding or buck hutch.

(See chapter on hutches.)

RUNNING-ON HUTCHES

When producing rabbits for meat purposes only, the young are usually weaned and kept together until killed. But when producing for fur and meat, with the intention of producing good skins, they must be grown for a longer period before the skins are in prime condition.

For that purpose, single hutches are necessary after the 4 months stage, otherwise they will fight and badly damage fur and health, especially the bucks.

Small hutches are all that is necessary for this purpose.

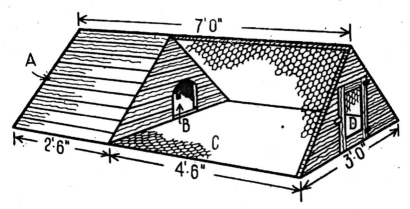

FIG. 32. APEX MORANT HUTCH FOR FOLDING

A Sleeping compartment. Weatherboarded or boarded and felted roof. Plain boards at bottom. Door at end, similar to D.
B Opening for rabbits to enter.
C Wire netting floor 1½ inch mesh. Outside covering of run 1 inch mesh.
D Door at end.
Length 7 feet, width 3 feet. Sleeping compartment 2 feet 6 inches long and run 4 feet 6 inches long.

An economical way is to make 8 hutches in one stack, the size of each compartment to contain one rabbit being 1 ft. 6 in. wide, 2 ft. 3 in. deep and 1 ft. 10 in. high. (*See Plate* 46, *Running-on hutches.*)

For running-on table rabbits up to killing age, small poultry houses with wooden floors are ideal. The approximate size required for an average litter of 6 to 7 is a house 6 ft. long by 2 ft. 6 in. wide.

MORANT HUTCH

A very useful system for growing young stock with the minimum of attention is the type of hutch known as the MORANT. This type is used for folding, the rabbits feeding through the wire netting of the bottom of the run.

This is particularly useful for keeping down the grass on lawns and waste grass in orchards. It is made to move easily, and generally needs moving on to fresh ground at least twice per day according to the number of rabbits put in and the quality of the grass being folded. Large plots of grass can thus be folded off in easy stages.

See the Morant Hutch (Fig. 32). It shows the apex type, which is easy to construct from new materials.

The type more often used is in the form of a coop and run. The bottom of the Morant run should be covered with $1\frac{1}{2}$-inch mesh netting. This prevents the rabbits escaping when folding uneven ground. The sides of the run are covered with 1-inch mesh.

Rabbits thrive well in these units and help to keep down the waste grass. It is important to note feeding is necessary in addition to the folding; grass supplies only part of the necessary diet.

In wet weather it is not advisable to move on to fresh ground so frequently. The same ground should only be folded once each season, and well limed before the next season.

COLONY HUTCH

Breeders who have several litters to wean at the same time, find advantages in a larger type of hutch for running on young stock.

Poultry houses can be used for this purpose also, the usual size being 10 ft. long, 7 ft. wide. Up to 30 rabbits can be weaned in a place of this size.

To keep trace of the records of stock so housed, if required for future breeding purposes, tattoo before allowing them to run together.

CONVERTED POULTRY HOUSE

There are many large poultry houses idle these days, which can very simply and easily be converted into most useful rabbit breeding and rearing units. See plan for converting poultry house design for partitions.

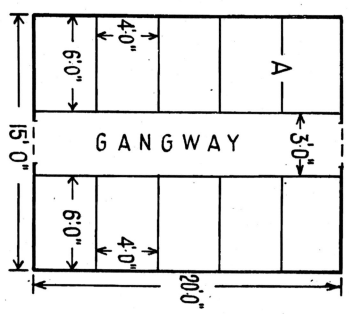

FIG. 33. CONVERTING POULTRY HOUSE

This house is 20 feet by 15 feet and is divided into Running-on pens for Rabbits.
Each pen constructed 6 feet long and 4 feet wide.
Gangway in centre. All doors opening on to gangway.
(*See Fig. 33A on dividing pens in large poultry house.*)

The house can be divided by having a gangway in centre if the house is wide enough, otherwise a gangway at side. Each pen should be 4 ft. wide, 6 ft. or more long, according to the size of the house. Partitions 3 ft. high.

These can be constructed of wood or wire netting, or both.

Feeding is done from over the top, but a door opening out on to the gangway is needed for cleaning out. The usual floor space per head for growing stock is 1½ sq. ft. Young stock grow exceptionally well in these runs, owing to exercise and good ventilation. Some excellent results have been achieved.

These pens can also be used for breeding in, allowing the doe a large sized nest box to have the litter in.

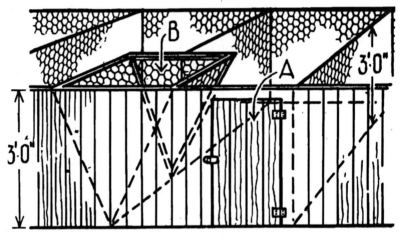

FIG. 33A. DIVIDING PENS IN LARGE POULTRY HOUSE

Height of all partitions 3 feet.
Width of pens 4 feet.
MATERIALS:—All wood or wood and wire netting ¾ inch to 1 inch mesh.
A Wooden door.
B Hay rack to serve two pens. Construct on wooden framing 2 inch by 1 inch, cover with 1¼ or 1½ inch heavy gauge wire netting.
(See Fig. 33 of conversion of poultry house 20 feet by 15 feet.)

HUTCH FLOORS

A sketch has been given on page 125 of a drainage system for hutch floors. This is specially illustrated because it is of the greatest importance. It is advised to make all hutch floors on this principle when possible. This system keeps the hutch floors clean and dry, and saves a lot of cleaning out.

NEST BOX

Many breeders favour nest boxes for the does to kindle in. These are placed inside the hutch at the darkest corner furthest away from the door.

The advantages are, (*a*) warmth, (*b*) keeps litter dry in damp bottom hutches, (*c*) prevents young from wandering about hutch too soon. The time when nest boxes are of the greatest advantage is during the severe winter weather, from November up to February. Also essential for breeding in open Colony runs. (*See* **Fig.** 34 *and specification of nest box.*)

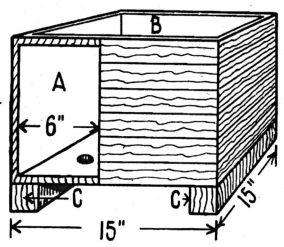

FIG. 34. NEST BOX FOR BREEDING DOES

Made from ordinary box similar to margarine box.
A Opening for mother to enter 6 inches wide.
B Top left open.
C Nail box on 2 inch by 2 inch quartering to keep off bottom of hutch. Bottom of nest box should have ¼ inch holes bored at corners for drainage.

WIRE NETTING FLOORS

Wire netting is often used for hutch floors with the idea of not using litter and improving sanitation. Generally speaking, for table and fur production, stock does not thrive so well as with the boarded floors. Sore hocks frequently develop. Large mesh netting prevents exercise and rabbits are liable to get their feet caught in it. Small mesh netting gets clogged with droppings, making cleaning out operations difficult.

These floors are most suited to Angoras, which need less exercise, and are purely a wool producing rabbit and not described in this book.

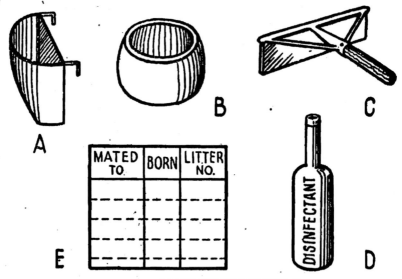

FIG. 35. HUTCH UTENSILS

A Water feeder. 5 inches by 3 inches by 2¼ inches.
B Feeding Pot. 4½ inches earthenware.
C Cleaning out hoe 6 inch or 9 inch blade.
D Disinfectant.
E Card for breeding record.

LITTER

Bedding litter is used in hutches to soak up any liquid and add to the general comfort of the rabbit. The best kind is wheat straw, next barley and oat straw. Straw, in addition to serving the purpose stated, provides a considerable amount of food of value. Another type of bedding which is good, and generally used by most rabbit breeders, is made by collecting waste grass from the hedgerows and roadside. Bracken, sawdust or peat-moss litter is satisfactory.

CLEANING OUT HUTCHES

How often a hutch should be cleaned out depends on size of hutch and number of rabbits in, and type of litter used. The usual period is once a week. Clear out all litter and droppings. Sprinkle bottom with lime. If any infection in hutch, limewash or creosote. When using the latter allow several days to dry before putting stock in.

UTENSILS

The following are the usual requirements as utensils for the hutch: (1) feeding pot; (2) water drinker; (3) hoe for cleaning out; (4) disinfectant; (5) hutch record card for recording matings, etc. (*See Fig. 35.*)

PROVIDING YOUR HUTCHES

Each breeder, doe or buck, needs a hutch to itself. In addition, you need at least two rearing-hutches to accommodate the youngsters between the time they leave their mothers and are killed for the table. You want two rearing-hutches because the sexes have to be separated at twelve weeks old.

The Requirements of the Modern Rabbit-hutch. The first point you must decide is whether the hutches are to be bought ready-made or are to be made at home.

If you are handy with your hands and are anxious to make a start with the smallest possible initial outlay, you will find it quite easy to build the necessary hutches. It is often possible to make use of old timber—odd lengths of flooring-boards and so on. Even more economically, you can use packing-cases and that without much alteration.

Let us have a clear understanding as to what constitutes a good hutch. There are three main points, all of which are equally important :

1. The hutch must be big enough to allow the rabbit to move about freely and to lie at full length. Also there must be ample headroom.

2. The hutch must be so constructed that it is capable of being easily cleaned.

3. The hutch must be soundly built and designed so that it is draught-proof but well ventilated ; if it

is to stand in the open it must be absolutely weather-proof.

The Simpler, the Better. Elaborately designed hutches are quite unnecessary. The simple one-compartment, open-fronted hutch is suitable for all purposes, including breeding. Does do not invariably make use of special breeding-compartments even when these are provided, and they complicate both the construction and cleaning out. You can always provide privacy for a doe at kindling time by fitting a wooden shutter to cover part of the open hutch front, or by hanging a sack there.

Shall it be Indoor or Outdoor Hutches? This depends upon your individual accommodation. If you have a spare shed you may be able conveniently to place your hutches inside. A vacant aviary might also make an excellent site for your hutches, or a potting-shed or a garage, even if it has a car in it. In the latter case, however, don't start up the car inside the garage ; the exhaust fumes may distress the rabbits.

A sheltered corner against a wall, provided with a lean-to roof, is another admirable site for hutches.

However, if the hutches have themselves to stand out in the open it doesn't matter. It only means that you must take particular care to see that they are of sounder and weatherproof construction.

You might suppose that rabbits reared in the open are hardier than those reared under cover. Actually, this doesn't seem to be the case. So you need not go to great pains to provide outdoor hutches when indoor are more convenient.

Essential Measurements of Hutches. You have learned that the first essential for a hutch is that it be large enough. The following minimum dimensions for rabbits of various sizes should be noted carefully :

> For rabbits of 7 lb. and over (Flemish Giants, Beverens, etc.), 3 ft. 6 in. by 2 ft. by 1 ft. 8 in. high.
>
> For rabbits of 5 lb. to 7 lb. (Havanas, Chinchillas, English, etc.), 2 ft. 6 in. by 2 ft. 6 in. by 1 ft. 8 in. high.
>
> For rabbits under 5 lb. (Dutch), 2 ft. by 2 ft. by 1 ft. 6 in. high.

These dimensions, you should bear in mind, are the *minimum* for the various types of rabbits. You should err on the right side by having your hutches too big rather than too small. Generally speaking, beginners are apt to use hutches too small for their stock.

" Packing-case " Hutches. A simple and efficient form of hutch can be made with only slight alteration from a bacon-box, a sugar-box or a match-case. The latter is as suitable as any. You may have to order boxes in advance from your local stores but the price will be only a few shillings apiece. The wood is of good thickness and the size, usually 3 ft. 2 in. by 2 ft. 4 in. by 2 ft. 2 in., is as near right as makes no difference.

Whatever packing-case you use, see that the wood is undamaged, the boards well jointed and the nails well home.

We have already said that single-compartment hutches serve quite well. If you prefer two compartments, partition off your box accordingly. The partition should have an opening through which the rabbit can pass, and should be fixed so that the sleeping-compartment is roughly one-third of the whole hutch.

With the one-compartment plan only one door is needed. With two compartments provide two doors, one solid, covering the sleeping-compartment, the other a frame covered with $\frac{1}{2}$-in. mesh wire-netting. The two doors should be hinged to either end of the hutch and may cover almost the entire front. A narrow strip of boarding nailed down each end of the front and where the doors meet at the partition makes a strong job of the front of the hutch.

The door of netting is made of a frame of light battens to which is fastened the netting ; it is a good plan to make two thin frameworks of just the same size, and to nail them together with the edges of the netting between them. By this means there are no sharp edges of wire to injure the animals or their owner.

Under the one-compartment plan, board up about one-third of the front, nail a narrow strip of wood down the other edge of the front and on this hang a door of wire-netting.

Some breeders having their hutches standing one on top of the other, in tiers, make a single door to cover all the hutches. The disadvantage of this is that rabbits in the lower hutches may hop out whilst one is attending to those in the upper hutch. That doesn't matter much when the hutches are in a shed—the runaways are easily captured—bu⁺
they are in the open it may lead to a troublesome chase.

Outdoor Hutches must be Weatherproof. When hutches are under cover no special precautions are needed to ensure that the inmates are not subject to dampness. When the hutches stand outdoors all necessary steps to make them rainproof will have to be taken.

Firstly, the roof should slope sharply, so that rain runs quickly off it ; secondly, it should project well beyond the front of the hutch to keep out driving rain ; thirdly, it should be covered with waterproofing material—roofing felt, rubberoid or even corrugated iron.

Building Better Hutches at Home. The rough hutches already described, adapted from boxes, are perfectly efficient and meet the needs of the meat-producer. They may be used to house the breeding rabbits or batches of weaned youngsters. But you may wish to construct rather more substantial hutches.

Materials needed to build a block of four hutches each 2 ft. 6 in. by 2 ft. 6 in. by 1 ft. 8 in. (high) raised 2 ft. above ground level on " legs " are :

> 6 uprights, 1½ in. by 1½ in. quartering, each 7 ft. long, for front and back framework.
>
> 6 cross-pieces, 2 in. by 1 in. batten, each 2 ft. 6 in. long, for side framework.
>
> 20 6-in. boards, each 3 ft. 4 in. long, for back and sides.
>
> 30 6-in. boards, each 2 ft. 6 in. long, for floors and top.
>
> 8 5-in. boards, each 2 ft. 6 in. long, for partitions.
>
> 4 pieces, 2 in. by 1 in. battens, each 3 ft. 4 in. long, for framework of doors.
>
> 4 pieces, 2 in. by 1 in. battens, each 2 ft. 6 in. long, for doors (top and bottom).
>
> 2½ yards, ½-in. mesh, wire-netting, 2 ft. 6 in. wide.
>
> 4 hinges ; 2 button fasteners, and nails, staples and screws.

The cost of the materials will depend upon the wood used. It is best to use new wood for the framework and doors, and the necessary battens and quartering will probably cost you ¾d. and ½d. per foot respectively.

For the boarding, good second-hand boards may do. The boarding for outdoor hutches should be ¾ in. to ⅝ in. thick ; for hutches to be kept under shelter you can use any light wood available.

First, make the framework and nail on boards for back, sides and top. Next put in floors and partitions.

Then make the batten frames for the doors, fix wire-netting, screw on hinges and put the doors in place, finally adding buttons to fasten.

Be careful of these fasteners. Rabbits are mischievous creatures and delight to play with anything loose. Unless door buttons are secure they may be nosed open, to allow the escape of the rabbits. For the same reason it is better to use steel hinges than home-made leather ones.

Home-made hutches can well be finished with creosote or some other reliable preservative, both inside and out.

Roofs for Outdoor Hutches. The specification given is for a flat-topped hutch to stand under shelter. For outdoor hutches, as already mentioned, tops should be made with a backward slope and a good overhang, back and front. In this case the back measurement should be 1 ft. 6 in. as against 1 ft. 8 in. in front, which means a corresponding reduction in the length of the back upright.

Make your roof boards 3 ft. long to allow of a 3 in. overhang at either end. Roofs of outdoor hutches should, as stated, be finished off with a good felt or other waterproof covering.

A felt roof cannot be beaten for appearance, and if the felt is renewed periodically, the wood being given a coat of tar at the same time, it will remain sound for years.

Galvanised iron or zinc is less expensive material, but with this a hutch is likely to be unduly hot in summer and cold in winter, and is liable to " sweat " on the under surface, which may lead to dampness. You can overcome these disabilities, however, by placing the galvanised sheets over an under-roof of wood. Straw packed between the wood and the metal will absorb moisture and also help to keep the atmosphere more temperate.

BATTERY HOUSES SAVE TIME AND LABOUR

As you have read, hutches may be made quite satisfactorily from wooden boxes, or simply constructed of bought timber. Such hutches fulfil most requirements and are reasonably durable if carefully used, suiting admirably where the aim is merely to breed rabbits for home use to supplement the wartime ration of meat.

If you propose going in for breeding on a larger scale, however, with a view to sales of rabbits, either wholesale or retail (and this, it may be said, is a way in which numbers of people could be of very great service to the country at the present time), you must arrange your hutch accommodation on a more elaborate plan. The battery system is here to be commended.

In this system, hutches are built in blocks, each consisting of a number of hutches arranged in tiers. The method is particularly economical of material as the floors of one tier form the roofs of the tier below ; and each partition serves as a side wall for adjacent hutches.

The hutches are usually housed in a shed but may stand outdoors if suitably weather-proofed.

Advantages of Batteries. The uniformity of the hutches simplifies your routine work and gives the rabbitry a neat appearance. Another advantage is that as one door is made to cover three or more hutches, according to the number of tiers, considerable time is saved when a number of rabbits have to be fed.

With a battery of eighteen compartments, each 2 ft. 6 in. by 2 ft. 6 in. by 1 ft. 8 in. high, you provide hutches suitable for housing medium-sized rabbits—those of 5–7 lb. weight—singly. You can also use them as breeding-hutches for large does and for running young stock in pairs, from weaning until they are ready to kill. The total area of such a battery is 15 ft. long and 6 ft. high overall, the bottom tier of hutches being raised 1 ft. from the ground on " legs," back and front. You can, of course, modify these measurements for construction of batteries with hutches of any size.

Materials required for Battery Building. For the battery mentioned you will require the following material :

14 uprights (each 6 ft.) of 1½ in. by 1½ in. timber.
8 cross-pieces (each 15 ft.) of 2 in. by 1 in. battening (for framework).
40 pieces (each 5 ft. long by 6 in. wide) of ⅝ in. matching (for ends and partitions).
90 pieces (each 2 ft. 6 in. long by 6 in. wide) of ⅝ in. matching (for floors).
12 pieces (each 5 ft. long) of 2 in. by 1 in. battening.
12 pieces (each 2 ft. 6 in. long) of 2 in. by 1 in. battening (for door frames).
18 pieces (each 2 ft. 6 in. long by 4 in. wide) of rough boarding (cross-pieces for doors).
30 pieces (each 2 ft. 6 in. long by 6 in. wide) of ⅝-in. matching (for roof).
Nails, screws, hinges and wire-netting (½-in. mesh).

Constructing the Battery. Now for the actual construction.

First make the framework (back, front and two ends) and screw it together. Next put in the partitions for the first set of three hutches, then the floors, nailing the boards to the framework.

You now do the same for the second set of hutches, and so on to the end. Now you nail on boards for the ends, back and roof—and the battery is complete with the exception of the doors.

The six doors, each one covering three hutches, consist of 5 ft. by 2 ft. 6 in. frames of 2 in. by 1 in. battens, with three cross-pieces of 4 in. rough boarding, one at the floor level of each hutch. The 4 in. depth here allows you to tack up cards giving particulars of breeding or other details regarding the inmates of the hutches, which will be found very useful.

You should pull the wire-netting of the doors taut, and fix it with staples to the inner side of the frames ; then affix the hinges.

The doors are now ready to put in position.

A simple button fastening may be made of 2 in. by 1 in. wood, screwed to the upright, or the doors may be fastened with cabin hooks or metal buttons.

How the Roofs are Fitted. You must make the roofs of all outdoor batteries with a backward slope, having a good overhang, back and front, to carry off the wet, and they must also be covered with waterproof material to ensure their being quite weatherproof. If you make the front overhang about 3 ft. it will afford you protection when feeding the rabbits in wet weather.

Protection for the open fronts of the hutches in case of necessity can be provided by sacking curtains or shutters of three-ply wood. The latter should carry three ventilation holes, each an inch wide, placed near the top.

THE FOLD-HUTCH SYSTEM IS POPULAR

Another method for those running rabbits in a large way is that known as the fold system.

What is a Fold Hutch? In the fold hutch (or Morant hutch, to give it its other name) the rabbits are housed on grassland, the floors of the hutch having wire-netting to provide for grazing. In this way the stock is allowed to develop on nearly natural lines, the hutch being moved to a new patch of pasture every day.

You can only use this excellent system of rabbit-keeping, of course, if you have sufficient grassland available. A large lawn may serve admirably.

The Fold Hutch's Advantages. One of the big points in favour of this system is that it reduces labour very considerably. On good herbage no hand-feeding is necessary, except in spells of drought, from spring to early autumn. By simply moving the hutch to fresh ground the rabbits are fed for the day, and their hutches, in effect, are cleaned out.

Another advantage is that the rabbit-manure, left when the hutch is moved, is beneficial to the ground.

If you use an orchard, for example, for grazing your rabbits with fold hutches, the trees continue to bear fruit, the land is manured, and the grass left for hay can, in turn, be cut for many feeding purposes.

Overgrown plots on holdings can be occupied temporarily by rabbits in fold hutches, with the object of getting the ground cleared of weeds while enriching the soil at the same time. Grass strips alongside country or suburban roads will also often afford pasture.

Management with the Fold Hutch. In places which are not exposed or bleak, and where the soil is well drained, you can leave rabbits out in fold hutches all the year round, shutters against the hutches keeping out cold wind and rain.

Rabbits, however, are never placed in fold hutches for the first time before mild spring weather has really set in. Their previous feeding should have got them accustomed to plenty of greenfood, for now their principal food will be grass.

You should feed additional greenfood when it is

available. Put it in the part of the hutch which covers the grass, of course, not in the protected portion. Hay should also be provided. If you want to give the rabbits any mash or grain, it can be placed in saucers or pots. This matter of feeding is fully dealt with in a later section of this book.

Be careful never to move a fold hutch on to frosty grass.

Building Fold Hutches. A good type of fold hutch, suitable either for use as a breeding-hutch or to house a batch of growers or adult does resting between litters, is one 6 ft. long by 3 ft. wide with a sloping roof, making the height 2 ft. in front and 1 ft. 8 in. at the back.

This size will take a doe and her litter, a couple of adult does, or four youngsters under four months.

Stud bucks at grass, for which it is wise to provide plenty of room for exercise, may be housed in hutches 5 ft. long by 2 ft. wide, the height measurement being the same as in the larger hutches.

In the larger hutches, 2 ft. of the total length should be occupied by sleeping or breeding quarters, which must be dry and have a wooden floor raised above the ground. This part of the hutch is fully boarded in at back, front and sides (apart from an entrance-hole communicating to the other compartment), the remainder of the front being covered with $\frac{1}{2}$ in. mesh wire-netting.

The " run " section has its floor made of $1\frac{1}{2}$ in. netting, tightly stretched round the bottom and stapled to the sides.

Shutters giving protection to the run, when necessary, are provided. They are made to slide in and out and have ventilation-holes near the top.

The hutches mentioned for stud bucks may be simpler than the above. In these the whole of the front may be open and all the floor of netting. In that event a wooden box is stood on its side at one end of the hutch, the top making a shelf for the buck to lie on and the box itself giving him shelter in rough weather. Before a doe is put into the buck's fold hutch for mating the box would, of course, be removed.

Fold hutches have no doors; access to the interior being gained through the roof, which is hinged at the back.

In all these hutches the top of the framework, back and front, is prolonged 6 in. beyond the length of the hutch itself to form handles by which the hutch may be moved by lifting.

Materials for construction are : framing (for body of hutch), $1\frac{1}{2}$ in. by $1\frac{1}{2}$ in. ; for roof, 2 in. by 1 in. ; boarding, $\frac{3}{4}$ in. matching, or similar suitable wood. The floor of the sleeping-compartment is on ledges about 6 in. from the ground level, making a step-down into the run.

INSTALLING THE RABBITS

Now our hutches are built and our rabbits have arrived, so our purchases can be installed in their new home.

First, however, see that everything really *is* ready for them. Assure yourself that the hutches are bone-dry inside. Damp quarters can cause a lot of trouble. The hutches may have been lying about in the open between the time of building and the delivery of the rabbits. If so, they are almost certain to be damp and will need drying in the kitchen.

Assure yourself also that the hutches, if they are to stand outdoors, are in the most favourable position available.

The best position for outdoor hutches is against a wall or fence with protection from cold north and north-east winds. If they face full south, so much the better. There should be a few inches of space between hutches and wall or fence to allow a current of air to circulate around them. Also the lowest hutches should not rest direct on the ground but on a few bricks.

The hutch floors must have some covering. Use sawdust if you can get it. It is a disinfectant and a deodoriser, discourages insect pests and soaks up the urine. The floors of hutches are preserved by its use and cleaning out is much facilitated.

Both hay and straw make bad litter. If you can't get sawdust, use dry earth—dry it in the sun or in an oven—mixed with ashes from the grate.

Don't cover the whole of the floor of the hutch with sawdust or earth. Put a layer an inch thick all over the sleeping compartment but for the rest of the hutch a good handful placed in a corner will suffice, for rabbits, alone among domestic animals, use one place only for sanitary purposes and keep the remainder of their living quarters dry and clean.

SOME MANAGEMENT HINTS

Cleaning Out. Hutches should be cleaned out at least once a week and in summer twice or three times a week may be necessary. Though rabbits are less trouble to look after than almost any other livestock, some smell may arise from their hutches if they are neglected.

For the cleaning-out job you will need a scraper and small shovel. With these you can quickly remove the manure and soiled litter. The scraper you can make at home from a small piece of tin or sheet-iron, or you can buy one, triangular in shape, for a few pence.

Rabbit manure is very valuable. It should be saved, dried, pulverised, mixed with dry soil and used as a fertiliser for vegetables and flowers.

When Rabbits gnaw their Hutches. Some rabbits, especially does, develop the annoying habit of gnawing their hutches. If twigs of deciduous trees—those that shed their leaves—are given to the offenders to keep them occupied, they will almost invariably cease the practice. If gnawing still persists, there is nothing for it but to cover vulnerable spots with sheet tin or close-mesh wire-netting.

Provide Shade in Hot Weather. Normally rabbits enjoy sunshine, but in very hot weather it is necessary to take precautions against troubles arising from over-heating, which means keeping the atmosphere of the hutches as cool as possible.

When the hutches are inside a barn-type shelter this trouble from heat is not so likely to arise, but when hutches are built facing south, especially when they back on a wall, you must, during anything like a heat-wave, prevent the fierce rays of the sun beating in and making the place intolerably hot.

You can easily do this by fixing curtains of thick sacking along the whole length of the front of an open shelter, or across open windows and doors of a shed that catches the sun. Damping these with water at intervals, or, better still, damping them with dis-infectant, will do much to keep the atmosphere cool and fresh.

If you spray the interior of the shed or hutches with disinfectant also, the whole place is kept sweet.

With outdoor hutches, you cannot cover the front of each hutch, as this excludes air and makes matters worse rather than better. If you can't move the hutches into a more shady place, try erecting an awning and curtains, formed of sacking tacked to bean-poles, all round the hutches except on the side facing north.

Extra Protection is needed in Cold Weather. Where the hutches are under cover, no special pre-cautions are required even during the most bitter spells ; there is no need, for instance, to bring an oil-stove into the rabbit shed.

With outdoor hutches even there must be no coddling in winter. All you need do is to screen the hutch fronts against driving snow or rain either with sacking or boards in which ventilation holes have been drilled.

If the hutches normally face north, north-east or east, it may be well to face them in another direction during the worst of the winter.

Housing

Rabbits are adaptable animals but adequate ventilation and floor space for exercise must be provided, together with protection against rats and other enemies, rain, draughts and extremes of heat and cold. They are usually housed in hutches, either indoor or outdoor. Specially constructed houses may be used for large studs, but in the normal rabbitry, housing may consist of simple sheds or outdoor hutches placed in a sheltered position. Young table rabbits may be kept in colonies but it is usual to keep breeding does and their young in individual hutches (see Plate III).

HUTCHES

Hutches should be designed to meet the floor space requirements of adult rabbits and, as a rough guide, 1 square foot of floor space should be allowed per lb. adult live weight. The minimum size should be 2½ feet wide, 2 feet deep (back to front) and 1½ feet high for small breeds, but for most of the fur breeds and large and medium varieties, the general purpose hutch, 3 feet wide, 2¼ feet deep and 1½-1¾ feet high, is suitable for a breeding doe, but when her litter leaves the nest she needs a double hutch. Giants need even larger hutches.

The labour involved in caring for the stock is reduced when they are housed in well-designed hutches. All hutches should be constructed so as to facilitate easy feeding and cleaning ; complete protection against rats should be given. They should be placed in a quiet position so as to avoid disturbance to the stock, which breed better when kept under quiet conditions.

Outdoor Hutches should be placed backing a wall or in a sheltered site. They should be raised off the ground on legs and a sloping roof of waterproof material, such as corrugated iron, asbestos or roofing felt, should project for 4-6 inches over the front and rear so as to give protection against rain. One-third of the front should consist of strong ½-inch wire netting (outdoors) and the remainder should be boarded up to give protection against bad weather.

Indoor Hutches are preferable to outdoor ones in large rabbitries as they provide shelter for the stock and better working conditions for the attendant in bad weather. Any building that is dry, well ventilated and free from direct draughts is suitable. Disused stables, poultry houses and lean-to sheds are commonly used. At Harper Adams College a 90 feet × 20 feet army hut provided accommodation for the breeding does. The hutches (Plate I) were placed in rows, face to face, with 5-foot wide passages between each row ; wide passages facilitated easy cleaning. Light is provided by hopper windows in the walls opposite each passage. The lowest tier of hutches is raised 9 inches from the floor to keep it clear of rising damp, from vermin, and for convenience of working. The dual-purpose hutch (Plate II) is fitted with partition doors, hinged across the centre of the ceiling, to make two half-sized hutches when more cages are necessary for the growing stock later in the season. Pop holes do equally well.

Floors. A level floor is generally used in preference to a sloping one. Self-cleaning and draining devices can be useful if properly designed and constructed but these are somewhat expensive. Wire-netting floors ($\frac{1}{2}$- or $\frac{3}{4}$-inch 18-19 gauge) or slatted floors, over sliding metal manure trays, are sometimes used, as they help to prevent the stock from coming into direct contact with their faeces.

Woodwork should be treated regularly with a preservative such as creosote on the outside and distemper inside. Wooden floors should be given a good coating of hot tar, extending 4 inches up the walls, to make them watertight. The tar must always be allowed to set completely before the hutches are used.

Hutches may be identified with individual number plates of metal or wood, or by stencilled figures. It is also useful to provide hutch cards or plain postcards on each hutch for recording notes.

Nest-boxes or breeding compartments are usual in outdoor and indoor hutches. During the winter extra warmth can be provided by fixing a temporary board or sack over three-quarters of the door. Removable nest-boxes measuring 12 inches × 16 inches × 15 inches high, with one side open to within 5 inches of the floor, may be used.

HOUSING

THE first question which arises when the housing of rabbits is considered is, "Shall I house my rabbits inside or outside?"

If a building such as a garage, poultry house, garden hut or barn is available, it is a good idea to use it. Hutches which are to be used inside a shelter need not be quite so strong and well made as those which are to be used outside, because they will not have to stand up to such severe weather. If does are to kindle in the early part of the year, when severe frosts may be expected, it is less risky if they kindle in a hutch which is inside a shed. Kindling often takes quite a time and in severe weather some of the litter may be chilled to death before the doe puts them into the nest. If outside hutches are used for winter breeding purposes, the hutch fronts should be protected with sacking to guard against the cold.

Another point in favour of indoor hutches is that it is easier for the owner to attend to the rabbits in bad weather if he is under cover. If no shed is available a lean-to shed made with a galvanised roof will do quite well, especially if it faces South, and is well sheltered from cold winds.

If no shed of any kind can be constructed, there is no need to fear that the rabbits will do badly. Rabbits thrive well in outside hutches provided that the hutches are warm and weather proof. In fact, they are better outside than in a stuffy, badly aired shed. Sheds which house rabbits must always be well ventilated otherwise the rabbits will have perpetual colds.

In outside hutches there is not much risk of colds, or of the spread of infection.

Both inside and outside housing have their advantages and disadvantages, and it is up to each individual rabbit keeper to house his rabbits in the way that suits him best. One thing he must ensure, and that is that the rabbits have ample room, and that their hutches are dry and free from draughts. The tame rabbit is a hardy animal and can stand a great deal of cold, but damp and draughts will prove fatal to it.

People tend to make breeding hutches too small. Rabbits need room to move about, and a doe should be able to get away from her litter if she wants to, and the litter should have room to get the exercise they need for growth. The minimum size for a breeding hutch should be 3 ft. 6 ins. long, 2 ft. 3 ins. wide and from 1 ft. 10 ins. to 2 ft. high. It is most economical of space and building material to make the hutches in stacks of three tiers. It is sometimes possible to get bacon boxes, or sugar boxes in the correct size, and this means that the hutch is half made. If wood has to be bought, ¾ inch tongued and grooved boards for back, sides, floor, roof and front is best, and 2 in. × 1 in. quartering for the framework. Some ¾ inch mesh wire netting will be needed for the doors, and roofing felt for the roofs if the hutches are to be used out of doors. All hutches should be raised well off the ground, so that a free passage of air round them is assured and the hutches cannot become damp underneath. It also prevents mice and other vermin from making their nests under the hutches.

The framework of a stack of hutches should be made in four sections: back, front and two sides. A stack of three tiers can be made up of any number of separate hutches in units of threes. It can be three, six, nine, twelve and so on, and it is easy to add an additional three hutches when there is an increase of stock. When the framework has been made, make frames for the doors. It is quite a good plan to have a door which opens along the whole length of the hutch with a small wire netting door hinged on it. The netting door should take up about half the length of the hutch if the hutches are for outside use. For inside use the netting door can be made larger as this permits more ventilation.

It is a common fault in home-built hutches to make the door so that it fits badly, and is either too big or too small for the space it is to occupy. If it is too big there is continual trouble in shutting it, and if it is too small it not only admits wind and rain into an outdoor hutch, but also gives the rabbit added opportunity for hutch gnawing.

When making doors, measure carefully to see that they are an exact fit. This will save a great deal of time and trouble later on, and will mean that the rabbits are more securely housed.

When making breeding hutches, fix a shelf almost six inches wide, and about nine inches above the floor of the hutch, running from front to back of the hutch behind the wooden door. This will enable the doe to get away from her hungry litter if she wants to. Just before weaning time young rabbits often worry the doe unnecessarily for milk, and if she cannot get away from them for a time, she may turn savage. Such a shelf also gives the youngsters more room in the hutch when the doe is not occupying the floor space.

Another thing which is most useful in a breeding hutch is a hutch board. This is a board about three inches high, which fits just behind the door and remains in position when the door is opened. The board forms a barrier which prevents litter and bedding from falling on to the floor when the door is opened, and it also prevents the young rabbits from falling out. Young rabbits have a habit of crowding to the door when it is opened, particularly at feeding time, and if they are in the top hutch, a fall to the ground can be fatal.

The hutch board should not be a fixture, otherwise it will make cleaning out difficult. A slot to take the board can be made quite easily by nailing narrow battens on either side of the hutch, so that the board can be lifted out when the hutch needs cleaning.

Some breeders build their hutches with a floor which slopes slightly towards the back, and projects a little beyond the back of the hutch. This is to allow the moisture to drain out, and fall to the ground away from the back of the hutch. If the floors are solid and well built, there is no need to use a sloping floor.

If the hutch floors are well creosoted, and sanded while the creosote is still wet, this will help to keep them waterproof. The entire stack of hutches should be given one or two coats of creosote both inside and out. This helps to preserve the wood, and is a precaution against infection if old wood has been used for hutch building. The creosote must be given plenty of time to dry before the hutches are used.

If the hutches are for inside use, they need no extra roofing, but if they are to be used outside, the roof must be covered with roofing felt and made with a backward slope. To obtain this backward slope, reduce the length of the back uprights by about 3 inches. It is a good plan to make the roofs of outdoor hutches project about a yard beyond the front of the hutches. This serves the double purpose of sheltering the hutch from rain, and it also protects the rabbits' owner when feeding or attending to the animals in wet weather. A hay rack should be made at the side of the hutch about six inches from the floor. This is usually more satisfactory than a wire netting rack to hang on the door.

When finishing off the hutch make sure that there are no rough edges of wire netting anywhere on which the rabbits may damage themselves.

Smaller hutches for single rabbits can be made on the stack or battery system in the same way. A good sized hutch for a single rabbit is 2 ft. 6 ins. long by 2 ft. wide by 1 ft. 10 ins. high. This gives room for the rabbit to move about freely, and is a very suitable hutch for pelt rabbits which have to be housed separately.

Where rabbits are not intended for the production of prime pelts, they can be housed together in indoor colonies or runs. Rabbits do well on this system, because it gives them ample room and opportunity for exercise. Colonies of this type can be made in an outhouse or garage, or in a lean-to shed. The pens can be made with wire netting of 1 inch mesh, mounted on wood framing of 2 × 1 inch. The usual floor space allowed in a colony pen is 2 square feet for each rabbit, and it is not advisable to run more than 25 rabbits in one colony. This is because rabbits in colonies are rather easily frightened, and will stampede in a mad rush if a stranger is seen, or they are startled by an unusual noise. If there are too many rabbits together, these stampedes may cause one or two of them to be injured.

When making colonies in a shed or out-building it is necessary to see that the shed is vermin proof, and that there are no floor draughts. If the floor is made of concrete, plenty of straw should be used as litter, and it is an advantage to put a raised sleeping board at one end. Some breeders give the rabbits a large box to sleep in, particularly if the colony is in a lean-to shed, as this gives them warmer sleeping quarters. Hayracks can be made on each wire netting partition, and if there are several separate pens, one hay rack can be built to serve two pens.

Colonies of young rabbits may be made up of one litter, or of several litters of the same age. They will live together quite peacefully up to about four months old, but must then be separated into sexes. The does will run together until needed for breeding or for table purposes, but bucks must be housed separately after this age or they will fight.

Morant Hutches.—During late spring and summer and for a part of the autumn, rabbits can be housed out of doors in Morant Hutches. The Morant system is a method of raising rabbits in batches of six or eight in movable hutches which have wire netting floors. The hutches are placed on grass, and are moved daily. This does away with cleaning out, except for the small covered sleeping compartment which is quickly and easily dealt with, and to some extent with the labour of feeding. Rabbits cannot be expected to get all the food they require from grazing the grass available when the Morant hutch is moved, but they will get quite a big proportion of it. When rabbits are housed in Morant hutches feeding costs are very small, but it is essential that the grass the rabbits are getting is good grass containing a variety of herbage. In addition to what they can pick for themselves, rabbits need some hay daily, and a little mash two or three times a week. If the green food is sparse or of poor quality, they will need some wild green food, or garden greens in addition.

There are several possible designs for Morant hutches, but each has the same essentials. The hutch must be reasonably light so that it can be moved easily, and must be fitted with handles for ease in lifting. One part of the hutch must be covered in as a protection against the weather, and the covered in part can have a board floor raised slightly from the ground. This serves as sleeping quarters for the rabbits, as it is not advisable for them to sleep on the ground. Some breeders prefer to make the floor or the covered part of the hutch of netting and to put raised sleeping shelves round the side. Either way is satisfactory.

The Morant hutch itself should be about 7 ft. long, 3 ft. wide, and rise to an apex 2 ft. 9 ins. high. The covered part of the hutch should be about 3 ft. long. Four triangles should be constructed of 2 in. × 1 in. battens each 3 ft. long. A fifth triangle should be made and boarded in, with a hole cut for the rabbits to go in and out. The hutch should open completely at either end. The netting used should be 2 inch mesh at the bottom of the run, and 1 inch mesh for the sides. If roofing felt is used for the covered part of the hutch, the whole of the sides should be covered with netting.

Young rabbits can be put into a Morant hutch at eight weeks old. It is best to put them into the hutch for the first time on a mild sunny day, and as with indoor colonies it is best to sex them before they are put into the hutches.

If young pelt rabbits are put into Morant hutches it is best to provide the hutch with some sort of shade, for although sunshine and fresh air are good for the rabbits, too much sun will fade their coats. Pelt rabbits really need separate hutches at from four to five months.

The full use of Morant hutches means that there must be ample ground to work on, because it is not advisable to graze the same piece of ground too frequently. If it is used too often the ground will become sour, and will cause stomach upsets to the rabbits which feed upon it. Each plot of grass should be lightly dusted with lime when it is vacated, and a really good crop of grass should be allowed to grow on it before it is used again. The same patch of ground can be used two or three times in a season, according to how rapidly the grass grows again. Care should be taken to see that the ground used for Morant hutches has not been fouled by dogs, otherwise tape-worm infection may cause trouble.

Morants can be used with success for about six months of the year, and often when the weather is too bad to have them outside they can be used to house rabbits inside a shed; Morant raised rabbits are usually hardy and suffer from few ailments. They do not mind dry cold, but excessive wet and fog are the things to guard against. Such outside hutches must also be well protected against the rabbits' natural enemies, such as dogs, cats, weasels, stoats and rats.

Free Range.—Rabbits can be bred and reared successfully on a free range. Poultry pens can be converted for this purpose, but to prevent the rabbits from burrowing out of such a pen, a wire mat, at least 2 ft. wide, should be laid on the ground inside the fence. This will prevent any risk of escape, as rabbits nearly always begin to burrow as near the edge of their enclosure as they can get. The guard fence round a free range of this sort should be about 6 ft. high, to prevent other animals from getting in. Wire netting of 2 inch mesh will do for the main part of the fence, but this is not proof against stoats and weasels, so netting of a small mesh should be used at the bottom of the fence. Cats will not usually attempt to climb a fence of this height, but should any do so, a 12 inch strand of wire netting at right angles on the top of the fence will prevent them from getting to the rabbits. About 7 sq. ft. of space per rabbit is a liberal amount of room in free range colonies.

When making up a pen of this sort, it is usual to use one buck with about a dozen does. The adult does will live together quite peacefully once they become used to each other, but bucks can never be relied on not to fight. Some breeders supply the does with boxes, or underground shelters, in which to produce their litters. Most breeders, however, prefer to take the does out of the colony and allow them to kindle in hutches. The difficulty with this system is that it is impossible to be sure when the doe is due to kindle, so it is necessary to remove all the does to hutches from about the twenty-fifth day after the buck was put in. The does will need to be kept in the hutches for thirty-one days after being removed from the colony, if they have not kindled before, because it is always possible that mating took place on the last day before the doe was removed.

The young rabbits, with their mothers, should be allowed to remain in the breeding hutches until they are five or six weeks old, then they can go back to the free range colony if the weather is good.

This system makes it very difficult to keep a rabbit's pedigree, because the youngsters get very mixed up, and a doe will often suckle youngsters which do not belong to her. The only accurate way of marking the youngsters is by ringing or tattooing. If young rabbits are being bred for meat purposes this system is quite a good one, and saves labour, but it is not a good system for fur production, or for the production of pedigree breeding stock. Stock run in free range of this type do very well if they are well fed, but they do need additional food to what they can gather for themselves. Plenty of greens should be given, and also some mash and hay. The green food can be tipped out on the ground, but it is a good idea to have some sort of container for the hay, and certainly the mash must be fed from small troughs, otherwise it will be wasted.

Colony rabbits must be securely shut up at night in some type of shelter. This not only means that they will be safe during the night, but allows the breeder to keep his stock under regular observation. If the stock are checked when shut up at night, and also when let out in the morning, any ailing rabbit can be spotted and, if necessary, removed from the colony.

If any difficulty is experienced in "rounding up" the stock at night, this can be overcome quite easily by giving the rabbits their last feed just before it is time to shut them up. They soon get to know when it is feeding time and will go into the shelter quite readily in search of food.

1. THE HOUSE

1. **Hutches and Rabbitries.** All hutches must be kept DRY, CLEAN, AIRY BUT FREE FROM DRAUGHTS.

Sunlight is good for rabbits and cavies, but hutches must not be in the full glare of the sun in hot weather.

Hutches may be stacked close together in a well-ventilated shed. Such a shed would be called a RABBITRY. A rabbitry is best if it faces south.

The floor of a hutch is generally made of wood, though rabbits are quite comfortable on wire-netting in their day run. Underneath the wire-netting must be a waterproof shelf or tray to catch the droppings. The sleeping-place must always be floored with wood. Wooden floors should be made waterproof with pitch.

Hutches must be made safe from dogs either by keeping them well off the floor or by a door to the rabbitry. A door made of wire-netting on a wooden frame, hung just inside the ordinary wooden door, is useful in a rabbitry, as the wooden door can then be safely kept open for ventilation whenever necessary.

All wire-netting in hutches and rabbitries should be small enough to keep out mice. A useful size is half-inch mesh.

A range of hutches should be designed so that the large hutches can be divided to make two smaller ones. A sliding partition, or a partition with a door in it, will do this.

A rabbit of average size ought to have a hutch at least 3 feet long, 2 feet deep and 18 inches high.

All fastenings to doors must be safe. Some rabbits get artful at opening doors that are insecurely fastened. Doors may be hinged to open sideways (like ordinary cupboard doors), or to drop down, or to lift up. Each way has its advantages. If a solid wooden door opens sideways or upwards there should be a board just inside the door to keep young rabbits from tumbling out if the door is opened suddenly. This board can be fixed in slots so that it can be removed when cleaning the hutch.

Some people like to give a doe a nesting-box in which she may make her nest. Such a box is illustrated on page 16. It is placed inside the hutch when needed.

Portable runs, to stand on the grass, should have wire-netting floors. The rabbits can graze through the mesh but are prevented from burrowing. (See Morant system, page 7.)

2. DIFFERENT
   TYPES OF
   HUTCH

MORANT TYPE
OF HUTCH

4. ANOTHER
   DESIGN FOR
   A MORANT
   HUTCH

2. **Bedding.** On wooden floors bedding is necessary to keep the animal comfortable and clean. A layer of sawdust or of peatmoss, with wheat straw on top, makes quite the best bed of all, but this may be too expensive. Wheat straw alone makes the next best bed. Oat and barley straw are not so good. Hay is sometimes used, but for Angoras it is most unsuitable, as it gets into the wool and starts a mat. The floor, if of wood, can be made waterproof by pouring hot, melted pitch over it and working the pitch into the wood with a hot iron. Wirenetting is sometimes used for the day compartment of the hutch. This needs no bedding, but a waterproof tray must be provided underneath the netting. The bedding in a doe's sleeping compartment should not be of *long* straw if young ones are expected. They might get tangled up in it.

3. **Colonies.** Rabbits have been kept very successfully in flocks or colonies. The flocks may be of any size up to 200 or even more, though the smaller ones of 20 to 40 are probably best. They may be kept either in a big shed or in an open run on the grass with a shed for shelter.

Colonies of does (females) and bucks (males) have been kept, the bucks mating with the does within a week or so of being put together. The proportion is about 15 does to 1 buck. About 28 days after making up the colony the does are removed and put in individual hutches, where, shortly afterwards, the does will give birth to their litters. Those does that prove to be not in kindle[1] must be mated again.

A certain amount of quarrelling may take place between does run together and between bucks, but once the rabbits get accustomed to the sight and smell of each other this quarrelling is not serious.

Several does (up to six or so) and their litters may be put together when the young rabbits are about 3 weeks old. This forms a 'nursery colony.' (See Picture No. 9.) At the usual age of weaning the does can be taken from the nursery colony and put back together into an adult colony.

The young bucks and does should be separated at about 12 weeks old.

The feeding of rabbits in colonies need not be very different from their feeding when in hutches. More green food can perhaps be used and it can be scattered freely and frequently over the run, or, better still, placed in low racks made of wire-netting. Concentrated food (oats, bran, etc.) can be reserved for the nursing does and young rabbits.

[1] A doe in kindle is one that has been mated and is therefore carrying unborn rabbits inside her.

91

Rabbits have been kept in empty poultry runs in order to keep down the weeds. The poultry and the rabbits use a run alternately, with a rest of a week or two in which the grass grows again.

Angoras have been kept successfully in colonies without damage to the wool. The rain makes the rabbits look rather draggled for a time, but when the wool dries it looks as well as ever.

For further information see Book No. 10.

4. **Morant System.** This is the name given to the method of keeping rabbits in wire cages on the grass. The run has wire-netting on the bottom, top, and sides. At one end of it is the hutch in which the rabbit shelters. This is either built into the run or stands separately, inside or outside. When built in the whole affair suggests a poultry fold unit, and can easily be carried about. A change of ground is necessary at least once a day. (See page 5.)

The wire-netting must be small enough to prevent the young rabbits from getting through. They can squeeze through an astonishingly small hole.

The grass grows up through the wire-netting bottom, and the rabbits can graze quite easily. If the grass is plentiful the rabbits may be able to get all the green food they need, and will need in addition only fresh water and a little good hay.

Cavies can be kept equally well on this system.

5. **Manure.** The dung and litter from the hutches is very useful manure for the garden. One way of treating it is to make a heap under cover, mixing in an equal weight of garden soil. The heap should be kept tight, and if it is very dry a little water should be sprinkled over it. When the heap is fairly well rotted it is safe to add it to the ground, but if applied while fresh in large quantities to growing plants it may do them harm.

Housing

Housing resolves itself into three main types :

(*a*) (*i*) Breeding hutches.
      (*ii*) Hutches for Stud Bucks.

(*b*) Running-on Hutches (for fur varieties).

(*c*) Colony Hutches (meat only varieties and fur varieties up to 4 months of age).

Prior to going into any detailed particulars of the different types of houses, a few general remarks will not be out of place. It must be remembered that the rabbit is naturally an open-air animal. The domestic rabbit is no less partial to good fresh air than is its wild counterpart. Fresh air does not, however, include draughts, extreme cold or strong sunshine.

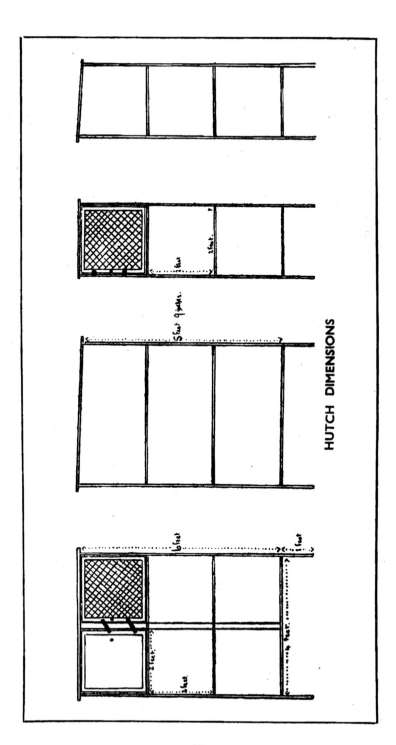

HUTCH DIMENSIONS

The best places for setting up hutches is in the open where fresh air can be enjoyed to the full, and also a good measure of light. Protection must be provided against inclement weather, and against strong sunshine striking directly on to the living quarters. Illustrations and drawings are given which show how these effects can be achieved.

Construction must be sound throughout. Good timber employed, and sound craftsmanship applied in the building. Roofs, walls, etc., to be free from cracks and poor joints. Doors should be slung on solid hinges, which will stand up to long use. Where hutches are arranged in tiers, every precaution must be taken to prevent drip from the hutch above into the one below it.

Having stressed the important points which are of general application, specifications will now be given for the different types of hutches suitable to the different uses to which they are to be employed.

Individual breeding hutches should conform to the following measurements :

Height, 2 feet ; Width, 4 feet ; Depth, 2 feet. Where hutches are built in tiers, the botton hutch should be raised 1 foot from the ground, and no more than three hutches should be placed one on top of the other.

The hutches are best equipped with a solid door covering half the frontage, and a wire mesh door covering the other half of the frontage.

Hutches for stud bucks will be found to be of ample size if built to the following dimensions :

Height, 2 feet ; Width, 2 feet ; Depth, 2 feet.

Running-on hutches need not be quite so large as those for stud bucks, and the following measurements will be found suitable for hutches utilised for running-on :

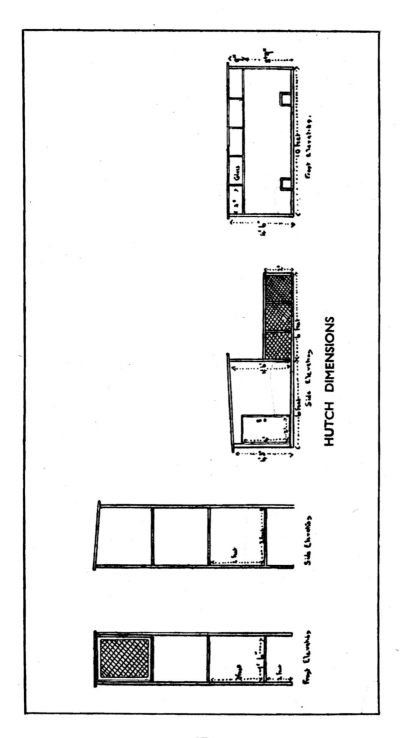

HUTCH DIMENSIONS

Height, 2 feet ; Width, 1 foot 6 inches ; Depth,
2 feet.

A house suitable for accommodating 24 or so young
rabbits in a colony from weaning to four months of age,
measures :

Height, 4 feet to 6 feet ; Width, 10 feet ; Depth,
6 feet.

A pen should be provided to each house, of similar
floor dimensions, where the young rabbits can take
their exercise. This should be constructed of half-inch
wire mesh, on three sides and the top, and join on to
the ends of the house. Young rabbits soon develop
burrowing propensities, and to avoid losses, the best
method is to peg small gauge, wire mesh on to the
floor of the pens ; 2 feet is quite sufficient height for
the sides of the wire-mesh pens.

Drawings and dimensions are given on pages 122,
123 and 125 for the various types of houses mentioned.

Housing.

Rabbits may be housed either indoors or out of doors. The cost of the housing is the same in each case, for when the former method is employed there is a shed in addition to the hutches to be provided, while in the latter much more substantial structures must be erected. The rabbits are equally healthy under cover as in the open ; hence the choice of methods is purely a matter of fancy. Under small farm conditions, however, the indoor method is advocated, since a suitable shed can most likely be found among the farm buildings which, with a little alteration, can be made to serve the purpose.

The shed must be wind- and wet-proof, well ventilated and fitted with windows, so that the interior is not too dark. The idea that rabbits do best in semi-darkness, although a very common one, is incorrect ; plenty of light with ample shading in sunny weather is the ideal. The selected shed should have a large window and a door on the south side and it should be not less than 6 ft. at its lowest part, so that the hutches may be attended to without the need for stooping.

Overcrowding is a serious fault in some rabbitries. There should be ample room for working at each hutch, while the fourth side of the house should be left free for corn bins, food receptacles and a bench or table for grooming purposes. The hutches may be arranged in three tiers, but the bottom one should be quite a foot from the floor so that it can be attended to easily.

Ventilation is an important factor in successful rabbit-keeping. Movable louvre-boards should be fitted in each end at the highest part, so that they may be opened and closed at will. This allows a larger sized ventilator to be used : a very useful thing in the summer.

Indoor hutches may be very simple structures, but for all that they should be well made. The best size for any of the dual-purpose breeds is 4½ ft. long, 20 in. high and 2 ft. from front to back. The sleeping quarters should be 2 ft. long, the run compartment the same length, while the remaining 6 in. serves for a hay rack. The front of the run should be filled in with ½-in. netting ; if a larger mesh be used the fur may be

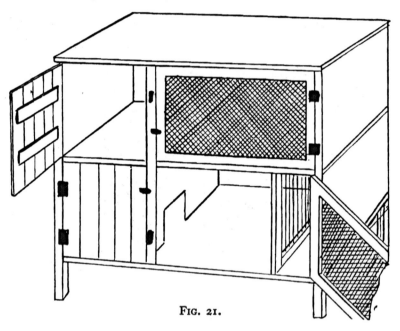

FIG. 21.

injured. An excellent type of indoor hutch is shown in Fig. 21. The timber should be ¾ in. thick, with 1-in. thick material for the floors. When the hutches are erected in tiers, the floors should be treated with tar to render them waterproof ; otherwise the urine may soak through to the lower compartments. When the tar is still wet the floor should be sanded.

Outdoor hutches should be built in three tiers and placed in the angle between a north and east wall with the front towards the south. A few inches of air space should be left between the walls and the back and end of the hutches. Appliances for outdoor use must be exceptionally well constructed of 1-in. thick timber on a stout framework of 2-in. by 1-in. battens. A sloping space roof should be fitted, this extending well over the walls, so that the rain is carried right away. A roller blind should be attached to the top of the

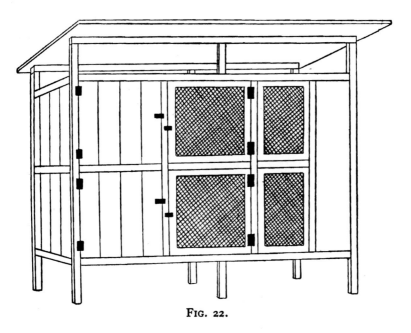

Fig. 22.

front, supported by iron rods at an angle of 45° when in use. A useful type of outdoor hutch is shown in Fig. 22.

With the area of ground available on the small farm housing the rabbits in portable hutches is an excellent plan. The rabbits can be left in the open whenever the weather is suitable and brought under cover when the conditions are unfavourable. A double portable hutch is shown in Fig. 23.

BEFORE coming to the more usual ways of keeping rabbits, we will refer briefly to the rabbit-court, which is a very profitable system of keeping hardy breeds, though it will not pay for Lops and other varieties requiring especial care, a question we shall deal with anon. Several rabbit-courts in this country are uncovered, but we believe such have not been found to pay very well, owing to the very wet seasons we are liable to. Covered ones are much preferable, and one of this kind we will now describe, with a few hints as to the best means of making it. If you have only " Hobson's choice," of course you must do the best you can ; but if you have a selection of ground, select a corner built against by two houses, and having the south side open. By this means your labour will be greatly curtailed, and if a third side is also built against so much the better. Dig out your foundations for the sides not already hemmed in by bricks and mortar, and go at least two feet below the surface. Lay your bricks nine inches thick—and it will be cheaper to employ a practical man for this part of the work—and let the wal' six feet high from inside at least. When this is com ..ed, have the floor well laid. If it is on soft ground it will be best to

THE PRACTICAL RABBIT KEEPER.

put down flags. *Ce n'est que le premier pas qui coûte*, and if a saving is effected by having an inferior article, a constant expense will be incurred in keeping in good repair. If the ground is clay, or hard, a cement floor will do better, and be cheaper. If this be laid, the first thing to do is to level the ground carefully, and then cut a ditch six inches deep down one side, sloping to a corner, where an aperture should be left, carefully secured by iron bars about two inches apart. Then mix some Portland cement with sand—you can ascertain the proportions on purchasing—and mix the whole into the consistency of a very thick cream. This done, lay the mixture on the floor, taking care to smooth very carefully with a board, and do it quickly, as the cement soon dries, and when dry it is as hard as a rock. Care should be taken to have the ditch or drain covered the same thickness as the rest of the floor, so that the depression is still apparent, as this is most essential for cleaning and scouring out the place. Further assistance will be provided by allowing the floor to slope very slightly towards this. A door must be made in the usual way, and a sliding board—such as is used for keeping young children within bounds—should be adopted, so that when the door is opened, any loose bunnies may not so readily make their exit.

It is necessary next to make hutches of some sort or other round the sides. We shall describe a few pages further on hutches of various descriptions, but for a rabbit-court a set of apartments as described here will be not only sufficient, but will answer better than any other. The side opposite the door, or facing the person entering, can be first dealt with. Make a front of inch board for the whole length, about 18 inches high, and fix it on the ground nearly two feet from the wall. Before doing so, make an aperture every two or three feet, according to the length of the place and the size of the proposed hutches, say two feet in an average court. The holes should be circular, and the edges should be lined with tin, or

103

the rabbits will gnaw them out of all shape. This done, divide the space into boxes, or hutches, with a hole for each leading into the open court. Cover this with inch board ; but let nine inches in width be fixed with hinges, so as to act as a door. One door for the whole is a clumsy way of doing the work, and it is not much more trouble to make one for each hutch, as in Fig. 1. This, opening as it does on the top of the hutches, is

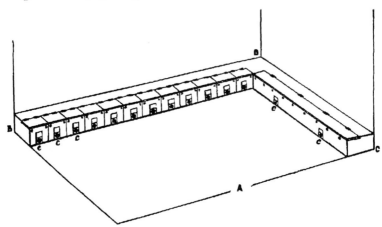

Fig. 1.—RABBIT-COURT.

A, Door to Court ; B B, Does' Hutches, with c c, Entrance Holes, fitted with trap doors ; B C, Sleeping Place for young Rabbits, with Entrance Holes and traps, c c. All the hutches and sleeping places have ventilating holes near the top, and hinge-flap covers.

very convenient for getting at the inmates, and for cleaning and feeding purposes. Each hole should be fitted with a sliding-door that can be closed at will, and each door should be fitted with a nut or bolt to fasten it down. These hutches will do very well for breeding hardy "table" kinds of rabbits, though they would not do for delicate breeds. According to the number you intend keeping, so large must be the accommodation provided for the young rabbits. A row along one of the other sides will generally be found sufficient. This need not be divided into hutches, and there need only be two or three holes cut, the whole forming a sleeping place which, with the timely addition of straw, will make all comfortable.

A mound of earth in the centre of the court is some times recommended, and it undoubtedly affords the rabbits lots of amusement. Still it has its corresponding draw-backs, and is objectionable on the whole. It often falls in, and buries the frolickers, and as it cannot be properly cleansed, the dung and urine saturate it, and give out a nasty odour that is anything but agreeable or healthy. A good layer of clean sand, changed twice a week, is better. The feeding-trough is a matter of importance, which will be dealt with later on. Fig. 1 is a rabbit-court floor with the hutches complete. A tile or slate roof is desirable, or a good serviceable one may be made of asphalted tarpaulin. Circum-stances must guide as to the construction, but in most instances a lean-to structure is found cheapest and most serviceable. It should be made to over-lay the walls of the court at least a foot, but the aperture between it and the top of the wall need not be filled in : indeed, ample ventilation in this or some other way must be provided. We have not gone into very minute details in these directions, but would refer the reader, in case of any difficulty, to our remarks on hutches further on.

The other method of keeping rabbits that we shall consider here is that usually adopted for breeding and rearing valuable fancy breeds, namely, in separate hutches in a rabbitry, or rabbit-house ; and as this treatise is likely to fall into the hands of many who do keep, or who intend to keep, fancy rabbits for exhibition purposes, we shall go into this matter more fully. Such a building as we have described would be unsuitable in many ways for a rabbitry, although, if it is necessary to make a commencement from the very foundation, the *modus operandi* would be very similar. The building would, however, require very much more finishing, and every aperture beyond the neces-sary arrangements for ventilation would have to be almost hermetically sealed, for, as we shall see as we advance, fancy rabbit keeping is a science of no little importance, and with a good deal of intricacy.

When you have finally decided to keep fancy rabbits, then —which you should not do without mature deliberation—look round for a spare room or building that you may have undisputed right and title to, and which will not be likely to be claimed for any other purpose for at least a reasonable time. Before giving directions for the making of rabbitries, we will briefly describe a few we have seen in use, and thus give some idea as to how existing edifices may be utilised for the purpose ; and as our readers will include many who have not much money to devote to the object, we will endeavour to show them how to effect their purpose without going to any great expense. The most peculiar rabbitry that ever came under our notice was a bed-room. Under most circumstances it is probable that such devoted to bunnies' use would very promptly lead to the desertion of all the other bed-rooms in the house, either to be also appropriated for the rabbits, or left vacant altogether, since rabbits do not smell particularly sweet, as a rule. The keeper, however, was so remarkably careful in his proceedings, and so scrupulous in his sanitary arrangements, that not only was the remainder of the house quite free from any unpleasant smell, but his bed-room itself smelt almost as sweet as any one could wish. The hutches were ranged round the room on benches about four feet from the ground, and along the backs ran a zinc tube sloping to a bucket, where the moisture congregated and was removed twice a day. Disinfectants were freely used, and the effect was quite satisfactory. Still, it is not a course we can recommend, as but few houses have occupants that would allow such a step to be taken, or sleep under the same roof as any member of the rabbit tribe, even when kept with the care that was noticeable in this case. A room on the ground floor would answer a great deal better, and would make a capital rabbitry provided there was sufficient space at hand. Another place we once saw requisitioned into a rabbitry, and whence several prize-winners came, was of a very different nature. It was an old wood-shed that had been emptied out and well cleaned, and every niche filled up, little or no ventilation being provided; the owner being a Lop-breeder of the old school, who believed that nothing but heat was required to procure perfection. The dimensions were, we should think, about 12 feet by 4 feet, the height being about 5 feet exclusive of the beams. When we had recovered from a violent bump against one of these useful but hard articles, and looked round, we were astonished to behold about fifty rabbits confined in hutches along the sides, and wondered how it was possible they could live under the circumstances. Indeed, the rabbits were but small, although many were very long in the ear. Still the heat from the paraffin lamp was most oppressive, and the animals seemed most uncomfortable. Not for a moment would we recommend any such course to be adopted. The age has passed when length of ear was the only characteristic in rabbits that called for attention ; good all-round property rabbits are now wanted, and such could never be reared in a cramped-up, badly-ventilated place. Another rabbitry we have seen was in a beer-cellar, and a very fair one it was. The place was roomy and well-ventilated, a little "beery" in its smell, certainly, and a little inclined to be chilly. To obviate this a gas-jet was kept burning all day, and the hutches used were of a warm sort. Another rabbitry was an outhouse neatly converted, and nicely ventilated, being heated with gas, and a trifle too warm for our liking. Still, it was a useful and, on the whole, decidedly good rabbitry, and demonstrated clearly the fact, that with care and a certain amount of skill, any existing building can be made into a very nice rabbitry.

Suppose, then, the would-be fancier determines to convert

an existing outhouse into a rabbitry. If it is very large it would be as well to appropriate a part only, and divide it from the remainder by means of a wooden partition, as a very large place is apt to be too cold for rabbits in the winter months. Having selected the building, or the portion of it you are going to utilise, see to the following matters the first thing, as without they are attended to all your subsequent labour will be quite useless :—

(1) See that there is no crevice or hole in the wall, as is often the case in disused outhouses. If there are any, stop them up. A brick building is far the best in many ways, and if there should be a hole in one of the walls there will be but little difficulty in filling it up. Mix a little mortar, or beg some from the nearest building in course of erection, and then plug up any hole with this and a small piece of brick or stone, smoothing the whole in a workman-like manner.

(2) Examine the roof, and see that it is quite water-tight. This need cause but little trouble, as heavy rain is pretty frequent, and then, if there is a weak point, Jupiter Pluvius may very safely be left to find it out. It is a more difficult matter to remedy any defect of this kind, but the cost of a labourer to do the work would be but trifling. This matter calls for imperative attention, as any leakage would do incalculable damage to the hutches and their inmates.

(3) See to the paving. If the floor is a brick one it will do very well, provided always it is in fair repair. We should not advocate the laying of a brick floor, because moisture is apt to sink in between the crevices of the bricks, and to cause an unpleasant smell. Even if there is a brick floor, and it is out of repair, a good and cheap way of rendering it A1 is to lay down a thin coating of cement, as already mentioned. Where the bricks are sound this will be very thin, but in the nicks and in any cavity the liquid will flow in and make all level. The time saved by this means in sweeping and cleaning after-

wards is very great indeed. If there is only an earth floo. level it well, get it stamped hard, and then lay your cement o it. Flags make a nice floor, but they are more expensiv generally, and certainly more trouble to lay properly.

(4) Look to the ventilation. Many a rabbitry has n ventilation at all, and the result is that whenever the door i open the foul air rushes out, giving the visitor a very bad in pression. Nature demands fresh air for everything in th animal and vegetable kingdom, and rabbits are by no means a exception to the rule. We have found it sufficient to leav small apertures under the eaves. Do not mistake draught fc ventilation. Let the current of fresh air flow high up in th room. This is quite sufficient, because the hot air is constantl rising, and is replaced by the cool and pure atmosphere.

(5) See that there is plenty of light. Except in very rar cases light is almost as essential for the well-being of rabbit as food and air. If there is no window, get one put in if yo can afford it. The cost will be repaid without a doubt. *A* good plan is to have two frames made, one of glass and th other of wire. At night, and during cold weather, let bot frames be down, but when it is very hot the glass can be raise either partially or entirely, as preferred. Have very secur fastenings both to window and door. Cats are clever thing: and can get through very small holes, but the depredations lai to their charge are often the result of the work of more accom plished but not less unscrupulous beings. Special provision will be necessary for Lop breeding, and these will be dealt wit] in their turn.

But supposing there is no place that can be adapted as it is, perhaps a shed open at the front can be found. A brick front and a door will not be very expensive additions, and the result may be a very nice rabbitry indeed. A wooden front will be cheaper, and if the boards are pretty thick, made to overlap one another slightly, and then tarred, the result will be fairly satisfactory. So also will be thick match-boarding. Still, the danger of fire is greatly enhanced, and bricks are for many other reasons decidedly preferable.

If there is no building, and no shed that can be adapted to a rabbitry, the work will of course be greatly increased. Still, for all that there is not so very much to do. Much trouble will be saved if a corner of two buildings or walls can be obtained, as two sides are then ready, and will be a firm beginning. Occasionally, such an angle possesses the double advantage of having a door ready made. If possible the other two sides should be made of brick, but we would not advise an amateur to do the building himself, as, although it looks very easy, it is a good deal more difficult to build than it appears to be. If expense must be avoided, a wooden frame may be made to suffice, in which case the following will be the simplest course to follow. Measure and mark out the size of your building, and at the corner facing the angle dig a hole about three feet deep. Purchase a strong stump about nine feet long, and something like a railway sleeper, and tar one end, which place in the hole, and fill in the soil, stamping it down firmly. From the top nail cross-pieces at right angles to the adjoining walls, thus completing the square or oblong. This should be fixed up very firmly, and the boards may be nailed across, overlapping, and tarred. A roof may be made in the same way, and may be so constructed as to be perfectly water and draught proof. Still there is great danger of fire if any heating apparatus is used, and the whole is not so snug or warm by any means.

Of whatever the outer walls are constructed, the interior should be limewashed before any hutches are put in, and this operation should be repeated once or twice a year, in which case there will be little or no fear of vermin. If the boards are painted, a better appearance may be got, but the wash is the healthiest and the best in many ways.

As to the size of the rabbitry, a good deal must depend upon circumstances, and the reader will be able to judge for himself. Height is a consideration, and a rabbitry should never measure less than 600 cubic feet of space, or be less than seven feet high in the centre. It will be well from the commencement to make arrangements for adequate cleaning out, for "cleanliness is next to godliness" in rabbit-keeping at any rate, and we would again refer our readers to the scheme for a gutter or channel to carry off superfluous moisture.

We now approach a most important section of our work, for it is to a great extent on the quality of the hutches that success in rabbit-keeping may depend. By laying before the notice of the reader a large number of hutches of different forms, with a few hints as to their construction, we shall put it in their power to make a suitable dwelling at a cost ranging from a few shillings to as many pounds. No matter of what the hutch is constructed, there are five indispensables, which are (1) space, (2) light and air, (3) absence from draught, (4) careful finishing, so as to prevent any roughness likely to injure the rabbit's coat, (5) durability. Without these five matters are attended to, all labour expended on a hutch is worse than useless.

In the previous pages we have given directions for the erection of rabbitries, or houses in which the hutches or cotes can be kept. Many fanciers have no such house at all, but for many reasons it is desirable, besides being essential for several varieties. But as the requirements of all will be studied here, we will give some directions as to the erection of outdoor hutches. These will do for Himalayans, Belgian Hares, and Patagonians, as well as those inside; and Silver-greys will be as hearty here as anywhere, though it is an admitted fact that the shading does not " come on " so rapidly as in indoor hutches. Dutch may be kept out of doors so far as hardiness is concerned, but when kept with other varieties, for reasons that will be discussed hereafter, it is advisable to keep them under

111

cover. Lops should not be kept out of doors. They require a little warmth and a uniform temperature, and these can only be obtained out of doors (and then but very imperfectly) by a process amounting almost to suffocation, and, it is hardly necessary to add, very unhealthy.

Dealing first with out-of-door hutches, or "outside" ones as they are generally called, the question of keeping out the elements is of paramount importance. When we first kept rabbits we had the hutches fixed up in a back yard, and every heavy rain drenched the inmates through and through. A cheap and excellent system for sheltering outside hutches, and one within the range of all, is to fix a shutter or old door against the wall or palings, and let it act as a lean-to roof. If carefully put up this will keep the rain off the hutches, which may be placed underneath it. But this is almost as much trouble as to knock up a rabbitry on the principle referred to in the last chapter, and it has the disadvantage of not being easily movable.

A nice stack of hutches, easily movable and fairly suitable for out-of-door use, may be made as follows :—Saw thirteen seven-foot lengths of nine-inch boards, an inch thick, four four-foot-six-inch lengths, and two three-foot lengths. This will make a good stout framework for a stack of four hutches. Four of the seven-foot lengths will make the back, three the top, three the bottom, and the remaining three the bottom of the top hutches, and the top of the bottom hutches. This last should be made very strong, and the joints tightly glued together to prevent moisture running through. For each side take a three-foot length and two lengths four feet six inches, placing the shorter in the centre. The three platforms will not have to be fixed at right angles with the back, but to slope as shown in Fig. 2, the object being to make the rain run off the roof, and to make the moisture run out at the back of the hutches. This work done, there will be a stout frame, and it will be noted

that the legs, A A, are formed by the ends of the side pieces. Th
next thing required will be to divide each of the compartment
into two, which can be done by nailing partitions exactly in. th
centre. The partitions will have to be cut as shown in Fig. ;
the top one being larger than that for the bottom, as is require

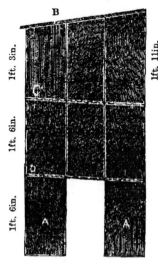

Fig. 2.

A, Supports formed by two outside
boards coming below the floor, D.
B, Roof, sloping and projecting at
front.
C D, Bottoms of hutches, sloping
to the back.

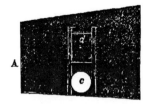

Fig. 3.

A, Partition between sleeping
and living room, in Top
Hutches.
B, Ditto, in Bottom Hutches.
c, Circular Hole, communicating.
d, Trap for ditto.

by the arrangement of the platforms. If these are inserted in
a workmanlike manner the framework will be considerably
strengthened, and four hutches will be made, minus the front.
The inside measurements of this will be as shown on Fig. 4.
Length of each hutch, three feet five inches ; height of top
hutches fifteen inches, of bottom hutches seventeen inches. The
next thing to do is to measure off spaces for sleeping compartments, which will be required for the breeding hutches.

Supposing you appropriate the two top ones for this purpose: measure off twelve inches on each side of the central partition, and divide the spaces thus obtained by two more partitions, making a circular aperture in each, as described a few pages back. In an outside hutch such as we are describing these are best fixed, and a little door either swinging or sliding along grooves over each aperture will often be found a great convenience. For each sleeping compartment make a solid wooden door in front, which hang on two hinges on the central partition, making them open back to back. This has somewhat of a left-handed appearance, but there is a gain in point of warmth.

For the front of day room a framework must be made. The size of this will be as in the figure on next page.* This is not a very easy task, but it can be done with a little care. For outside hutches the thickness of the framework should be three inches at least, while a little more than half that will do for those for indoors. Get iron bars, such as can easily be obtained from any ironmonger, varying in circumference from half an inch to an inch, and fix these in an upright position about an inch apart. This done, the frame is strong and durable. Fix it to a swing outwards with two strong hinges, and drive in staples for a padlock. Then the breeding hutches are about finished. For the hutches below, which are not going to be used for breeding, the fronts may be made of one piece, about half being made of solid wood, and the other of wires, as shown. A much easier plan is to nail three-quarter-inch wire-netting on the frames instead of the iron bars. For indoor hutches there is no reason why this should not do fairly well, but it is scarcely strong enough for outside ones. The hutch is now complete, and a reference to Fig. 4 will explain pretty accurately the construction of the front.

To make assurance doubly sure on the score of weather,

* It should be noted that these measurements will be affected by the different methods of joining the frame.

the top of a range of out-door hutches may be made to protrude about three inches over the front, and a shutter may
be fixed to the end by strong hinges, so that during the night
or during heavy rain it may be let down, while during the
day it may either be lifted up entirely or partially. Instead of
hinges hooks may be substituted, and the shutter taken off
altogether during the day. This is not shown in the figure, but
it is very simple in construction. It should be borne in mind

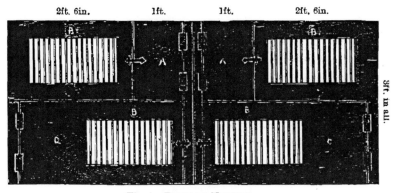

Fig. 4.—FRONT OF HUTCHES.

A A, Door of Sleeping Compartment for Does; B B, Open Frame, with wire or bars;
c c, Dark portion of Hutch, for Bucks or Young Rabbits.

that if the sloping roof comes out beyond the front, the top
doors will not open if they go to the top, and to obviate this
a slip of wood must be nailed along the whole top three or
four inches deep, and the top doors made correspondingly
smaller. The hutches as now finished will make a nice stack,
and will be standing eighteen inches from the ground, thus
preventing damp. The great drawback to a stack of this
kind is the wet running through from the top hutches to the
bottom ones, but we have found this not to occur when the
joints are very well constructed. Along the whole length of
the uppermost bottom, and at the extreme back, run in a zinc
groove or gutter, and let this slightly slope to one end of the

hutch, where it can protrude a couple of inches. Pursue a similar course with the other floor, and then when the rabbits are in, a bucket can easily be placed to catch the urine, which makes excellent manure, but which has a most offensive effluvium. With the exception of a few little sundries, which will be considered *seriatim*, everything is now complete, and as compact an outdoor set of hutches has been made as well can be. Presuming there is no better place, let it stand up against a wall, or in a corner, so that the wind may be sheltered. Always let the front face the south or west (the former is preferable), and, if possible, get a corner where there is some protection both from north and east; the winds from these quarters are most severe, and rain and snow are almost invariably beaten in from them, especially from the north-east.

Single hutches are not so well for out-door purposes, but if used the same principles may be worked upon, and they should be stood a short distance from the ground.

More important by far to most of our readers will be the subject we have next to enter upon—viz., indoor hutches and their fittings. As to the modes of construction, their name is legion. Lops being the most valuable specimens, we will consider their claims first, and describe a hutch that has been found highly successful for Lop breeding, and which is not remarkably common. We may commence by stating that for rabbitries single hutches are decidedly the best, saving a great deal of trouble, and being more easily kept sweet and clean. The special hutch alluded to is not generally made very large, the length being usually about three feet, breadth two feet, and height fifteen inches. We prefer a little larger, but will, for the purposes of illustration, adopt this size here. Make a strong box frame, having the inside measurement as stated. Appropriate one-third of this space at either end—the right is the usual, but it does not matter very much—for a sleeping compartment, which partition off in the manner already described. Instead

of fixing the partition as in the outside hutch, fit it in bet
grooves, so that it will slide out at will, thus enabling clea
to be carried on without so much difficulty.   For the sle
compartment make a solid wooden door, and fix it on in
usual way.   For the day room also make a solid wooden
but cut out a hole about nine inches square, and cover
with thin aviary-wire.   Next nail two wooden slabs o1
edges of the aperture, the slabs to taper from two inch
width at the top to nothing at the bottom.   The object

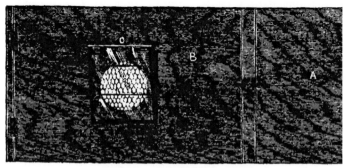

Fig. 5.—Close Hutch for Lops.

A, Door of Sleeping Compartment; B, Ditto of Living Room; c, Glass S
Light and Ventilation, sloping outwards towards top, which is open.

slip a piece of glass down, and thus keep in more
This can easily be done in one of three ways.   A groove
may be cut in the wood, down which the glass may slide;
strips of zinc, about an inch wide, and projecting inwards,
may be nailed on the edges of the slabs, and the glass allowed
to rest against them; or a wire may be stretched across from
slab to slab, an inch or two from the top, and a similar
distance from the bottom, thus providing, perhaps, the safest
plan of holding the glass that can be imagined.   The an-
nexed figure (Fig. 5) represents the work of the front, and
the glass fixed in position by the two wires.   This hutch
has been found to answer very well indeed with Lops; but we

117

do not on the whole recommend it. The time has come when it is no longer thought necessary to employ excessive or unhealthy heat, and there is not enough ventilation here to satisfy the minds of the majority of fanciers of the present day. But at the same time it is a good plan to adopt the system in ordinary hutches when a little extra warmth is required, and the slabs may easily be fitted upon the frame of any hutch. The benefit is, that the foul air naturally rises and goes out at the aperture of the glass, while sufficient fresh air will come in to be consistent with warmth. Glass being as well a bad conductor of heat, extra warmth is secured.

For the hardier varieties we should say, follow no such plan at all; but make a hutch a little larger if possible, and give plenty of air. Thus, instead of making the frame for the front room very thick and solid, it may be constructed as thin as possible compatible with strength, so that there shall be no limit to the amount of fresh air passing into the hutch at all times. The sleeping compartment may be made as already suggested, the partition being movable, and a sliding door being affixed to the aperture. Instead of one door to the sleeping compartment, we have seen recommended another plan, which we have tried with considerable success, and improved upon. It is very simple, the only difference being, that instead of having a door to fit the whole of the compartment, so fixed that when opened the whole is exposed, it is so arranged that a second door has to be opened before the nest can be disturbed, or the young ones fall on to the floor. It is done in this way :—After fitting the door on as usual, fix grooves just inside the door, and about six inches high. Now make a small slide to fit in, the length of the breadth of the compartment (twelve inches), and about six inches high. This can easily be slipped in, and its own weight will keep it in its proper position. Fig. 6 represents a hutch made as suggested for indoor purposes, with the second door for the sleeping

compartment, as sketched out, the outside door being op
The benefit is at once obvious. It is often necessary to b
a peep at the young rabbits before they have been in the w
many hours, and when a sudden exposure to cold may be fe
Moreover, the doe often makes her nest right against the d
so that if opened suddenly, the whole would fall out wit
rush, and the young be killed. If this plan is followed
the door can be opened, and there will still be a door half-'
up. This need not be removed ; and any necessary obse
tions may be made with more ease and safety. Another]

Fig. 6.—BREEDING HUTCH.

A, Front of Living Room; B, Door of Sleeping Compartment; C, Inner Slide, fi
half the door space when door is open.

is to substitute two doors for the compartment, by cutting
door in half, and using hinges to both. The objection is, tl
is more chance of draught, besides more trouble in the ʻ
struction.

Sometimes wooden bars are substituted for the iron o
their chief claim being the low price at which they can be
tained ; but in the long run they are much dearer, as the rab
gnaw them through, and they continually want renewing, iɪ
pendent of the dangers of broken legs by jumping through
space that may have been enlarged by the teeth. Iron hoops,
such as are used for barrels or tubs, may be used, and they do
very well, especially for outdoor hutches ; the best plan is to
interlace them and fasten firmly to the wooden frame. For
inside hutches, half-inch aviary-wire may be pretty generally

used, although many fanciers prefer the iron bars. The aviary-wire is much the cheaper, and a great deal of time is saved in cutting the bars the proper lengths, and fixing them in the frame. It is important in fixing aviary-wire to eschew tin tacks, and insist on being supplied with the staples made for the purpose.

Fig. 7 represents the front of a hutch on a principle we do not recommend, but which is very common, having the special advantage of being very easy to construct. A will be seen to be the door of the sleeping compartment, and B the door of the

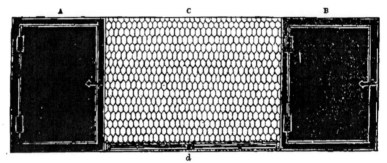

Fig. 7.—WIRE-FRONTED HUTCH.

living room; c is a wire front, which may consist of iron bars or wire netting, or, if preferred, after the caution given, wooden bars, which are so often jumped at because they are so cheap; d is a drawer for cleaning-out purposes, which we shall speak of further on, and which is absolutely necessary in hutches of this construction, unless the keeper is prepared to undergo a very large amount of unnecessary and very unprofitable labour. For persons who are very anxious to do things cheaply this hutch may suffice, and if it is impossible to get a little practical help in the construction, it is perhaps as well to have one like this for a start. Of course the frame must be made in the same way as previously recommended, but a wine or egg case may often be got for a mere song just about the correct size, and very neatly and firmly constructed. If one of these is bought, commence operations by a repeated scouring, and well limewash before bringing into use.

It seems to be a characteristic of the human race that they will persist in using things for purposes for which they were never intended, and hence it is that tubs and barrels are fre quently impressed into the service of a rabbit fancier. For many reasons they are highly objectionable and most inappro priate, especially when they are simply stood on end, and poor bunny placed in, having no other view of the outer world than he can obtain by star-gazing. It is impossible to clean out such an article without turning it over, and as this is a good deal of trouble it generally gets done about one-tenth as often as it should be. Of course, in cases of emergency it is all very well to pop a rabbit in a tub for a day or two; but to make one of them a permanent dwelling, every true fancier in the kingdom would at once say was preposterous.

There is another plan of hutches that has but one objec tion—it is that it requires so much room. When, however, there is plenty of this very useful and necessary commodity we should recommend its adoption. The wire front is fixed firmly to the body of the hutches, there being no framework door, and no opening from the front. This is supplied by means of the back being made to open as a large door, and thus the cleaning process can be very easily carried on. This is the very best plan that can be adopted when there is sufficient room, as hutches so made are very strong, and the difficulty in keeping a frame in good repair is avoided.

Passing from the consideration of the hutches themselves— and we think the matter has been so fully dealt with that there need be little difficulty in making what is required—we will refer briefly to a few of the most necessary fittings.

Cleanliness being very important, it is not very singular that many schemes have been mooted for the purpose of assisting the process of cleaning. One of the best is to cover the boarded floor with a double coat of paint, and while the second one is still wet, to dip it into a bath of Portland cement. This dries very hard indeed, and can be scraped again and again. It answers better than a simple cement floor, because this is colder and does not always stick well to the wood. This scheme is decidedly a good one, and is very simple indeed. As to expense, it will "pay for itself" ten times over in as many years by keeping the boards from rotting. Zinc floors we object to, on the ground that they are very cold to the rabbits' feet; otherwise they do well, and are very easily cleaned indeed. Slate slabs are open to a similar objection, besides their immense weight, which

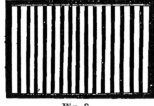

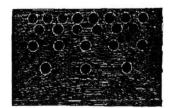

Fig. 8. Fig. 9.

often destroys a frame altogether after a few months. Either of these two latter may be used with success if a false floor is provided. There are two or three methods of making them, which we will briefly describe. Fig. 8 is a false bottom composed of wooden bars nailed on to cross-pieces. These should be very strong, while the bars themselves may be from one to two inches across and about half an inch apart, the edges being slightly rounded. The effect of this is that the inmates in moving about unconsciously shake all the droppings through on to the real bottom, from which they are easily removed, or on to a tray, which can be drawn out and washed at pleasure. The framework bottom must be exactly the same size as the living room. Care must be taken not to leave sufficient space between any two bars for a leg to slip down, or a surgical operation may become necessary. For the same reason it is obvious that the work must be very neatly done and the boards carefully planed. A preparation of paint and cement will be as well to make the bars more durable and less liable to be soiled. The space between the top and real floor should be about two inches, and this is easily secured by fixing the cross-pieces undermost.

Another plan is to have a second floor made of wood, and to bore holes about an inch in diameter in it, mostly towards the back, but with a few towards the front as well, as shown in Figure 9. These are made the centres of a number of slightly undulating valleys, down which the droppings and wet run, and thus on to the real floor. This is rather better than the

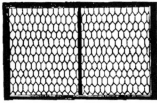

Fig. 10.

bar principle, because there is less danger of the rabbits getting their legs through them; but they are not generally in use. If properly constructed one will last as long as a hutch.

Fig. 10 represents a species of false bottom which, although very effective in keeping the rabbits clean and in allowing the droppings to fall through, is open to other objections. A frame is made the size of the bottom of the living room, with a cross-bar dividing it in two. Across the whole a wire netting is nailed, the cross-piece acting as a strengthener. The mesh should not be more than half-inch, or the rabbits' legs will get through and be in danger of getting hurt. If the hutch is very large two cross-pieces will be necessary. The objections to this plan will be seen at once. It acts remarkably well, but is most uncomfortable to the rabbits themselves, whose comfort should receive a certain amount of consideration. Their skin is very tender, and it is obvious that when lying on it they will suffer considerable inconvenience. Nor is this all. The wire is generally a little sharp at the joins, and in this case the feet of the rabbits get lacerated and torn, or, at any rate, made sore. Any rabbit keeper who has had much experience will be aware of the evils of sore hocks, and of the necessity of adopting means for preventing them.

There are other false bottoms in occasional use, but as they are on the same principle as those already described they need not be referred to. When a false bottom is used, to make matters perfect a zinc floor should be undermost, or, better still, a zinc movable tray, on the bird-cage principle, should be brought into requisition.

We have described these additions because they are greatly used, but while doing so we would add that they are quite unnecessary. Simplicity is always best; and the simplest rabbit-hutches are generally found the best for all purposes. The false floors are elaborate, but they cost a great deal of labour, take up a great deal of room, require the hutch to be 2 in. higher, detract a great deal from the comfort of the rabbits, save little or no trouble in cleaning, and do not keep the hutches any sweeter. A hard sound floor can be cleaned with as much ease and in as short a space of time as any elaborately-designed framework that can be imagined.

A few words as to fastenings will not be out of place. Undoubtedly, a lock and key on the outer door and a bolt to the sleeping compartment form the best. The bolt can be so fixed that it can only be propelled and repelled from the day hutch, and thus a second lock is dispensed with. When the rabbitry door is always kept shut a bolt will do very well for both doors; but whatever plan is adopted it is impossible to be too careful, as if the hutch door swings open the results may be very disastrous.